D0866151

Structural Steel Fabrication Practices

Structural Steel Fabrication Practices

John W. Shuster

McGraw-Hill, Inc.

New York   San Francisco   Washington, D.C.   Auckland   Bogotá
Caracas   Lisbon   London   Madrid   Mexico City   Milan
Montreal   New Delhi   San Juan   Singapore
Sydney   Tokyo   Toronto

Library of Congress Cataloging-in-Publication Data

Shuster, John W.
  Structural steel fabrication practices / John W. Shuster.
    p. cm.
    ISBN 0-07-057770-6 (hardcover)
    1. Building, Iron and steel. 2. Steel, Structural. I. Title
  TA684.S544 1997
  624.1'821—dc21   96-37443
                                            CIP

Sci
TA
684
S544
1997

McGraw-Hill

A Division of The McGraw·Hill Companies

1 2 3 4 5 6 7 8 9 0 BKP/BKP 9 0 2 1 0 9 8 7

ISBN 0-07-057770-6

*The sponsoring editor for this book was Zoe G. Foundatos, the editing
supervisor was Sally Glover, and the production supervisor was
Claire Stanley. It was set in Garamond.*

Printed and bound by Quebecor Book Press.

McGraw-Hill books are available at special quantity discounts to use as
premiums and sales promotions, or for use in corporate training
programs. For more information, please write to the Director of Special
Sales, McGraw-Hill, 11 West 19th Street, New York, NY 10011. Or
contact your local bookstore.

To my granddaughter, Robyn Michelle Andrews,
with love and appreciation.
Your computer expertise made it all possible.

Acknowledgments

I am very grateful to the following people and the firms they represent:
Kenneth J. Schmidt of General Trailer Company; Alan Babb of Delta Sand and Gravel Co. and Delta Construction Co.; Patrick B. Eagen of Farwest Steel Corporation; Don Faughn of Eugene Equipment Co.; D.J. Carney of United States Steel Corporation; Charles R. Fassinger, director of technical services for the American Welding Society; and Edward Humes and Carl Horstrip of Lane Community College. The technical materials and the photographs they have provided are a very important part of this book. Also, thanks to my loving wife, Catherine, for the assistance of providing the format for the photographs and artwork.

Contents

Introduction

I am very pleased that McGraw-Hill is publishing this work. It should provide the answers to your questions on steel fabrication practices. It explains what steel is, as well as the changes that take place in the fabrication and welding processes. It considers the types of major fabrication industries, because you may want to look at some techniques used in companion fields of endeavor. Explanations of how storage tanks and ships are constructed may not seem to affect your work, but the tools and techniques used can show you new ways to improve your skills.

This text provides tips on techniques and safety considerations and explains some mistakes in welding repair and how to correct them. Projects featured in the text are fairly simple, but they introduce the beginner to a hands-on approach to fabrication.

This book can be roughly divided into four parts: materials (Chapters 1 through 5), tools (Chapters 6 and 7), applications (Chapters 8 and 9), and processes (Chapters 10 through 12). The book also features a list of abbreviations and a glossary.

Chapter 1 discusses mild steel, its uses, and its properties. Chapter 2 covers shapes, tolerances, availability, and pricing of mild steel. Chapter 3 introduces low-alloy, high-strength steels and their uses and properties. In Chapter 4, we discuss the shapes, availability, and pricing of different grades of steel. Chapter 5 explains cold-roll steel, alloys, and high-carbon steel.

Chapter 6 describes hand tools, including measuring, layout, and marking tools; hammers; wrenches; welding and cutting equipment; and hoists, clamps, and other holding tools. Chapter 7 features production and power tools, including shears, brakes, iron workers, positioners, cranes, riveting guns, welding machines and torches, and cutting equipment.

Chapter 8 discusses forming, cambering, cutting, shearing, punching and drilling, aligning, and fitting steel and steel structural shapes, while Chapter 9 covers shop organization and procedures and safety considerations. Chapter 10 is about the fabrication of steel structures, including bridges, dams, storage tanks, trailers, and ships. Chapter 11 goes into detail about welding practices and procedures, and Chapter 12 covers testing, certification, and codes.

I hope this book will encourage you to think more about your own safety and that of those you work with. If you work smarter *and* harder, you will always succeed.

Project list

1. Basic metallurgy tests
2. Cambering
3. Structural shapes
4. Pipe joints
5. Square-to-square transition
6. Making patterns
7. A scale-model round tank
8. Utility trailer
9. Truck bumper
10. Welder qualification test

Structural Steel Fabrication Practices

1

Mild steel

I believe that if we are to work effectively in the field of steel fabrication, we must understand the makeup of the materials, their properties, and the reactions to outside forces. Many of the fabricators and welders in industry today are excellent craftspeople, but they have never had an opportunity to study the properties of the materials they use. The basics of grain structure and crystal formation play an important part in why and how a material reacts as it does. Now is the time to take a little closer look at some of the structural steels.

We should, at once, consider steel itself. It is, of course, smelted from iron ore and then turned over to rolling mills. The mills first *soak* their *ingots* (Fig. 1-1), bringing them to a rolling temperature of about 2200 degrees Fahrenheit (F). A *bloom* for rolling is generally square or rectangular and has a cross-sectional area of more than 36 inches. *Billets* are the same shape but have a smaller cross section. In most cases, a *slab* is wider and flatter than a billet.

The steel shapes that fabricators are familiar with all come from rolling. The rolls turn at the same speed in opposite directions, as shown in Fig. 1-2. This process is called necking down or drawing out.

The rolls need not be smooth but may have shaping flanges (Fig. 1-3). These rolls are used in banks, and the steel may be reversed back through a constant lengthening process. The rolling of steel makes it tougher and stronger. After rolling, the planes of grain structure fall along the axis of the steel, which simply means that the crystals are longer and thinner. They overlap in layers. A crack in the frame of a truck is never a straight line. The path of the fracture follows the line of least resistance. That path follows the outline of separate crystals as much as possible, only breaking these crystals when a detour is virtually impossible.

If we heat a piece of flat mild steel plate with an oxy-fuel gas torch, we often see beads of water appear on the surface. This water had been retained within the spaces between crystals. This space is where gasoline is retained in the empty gas tank that explodes when the gas is vaporized by a cutting or welding operation.

Chapter One

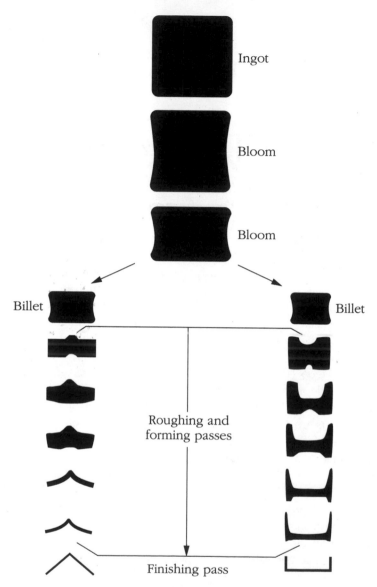

Ingot

Bloom

Bloom

Billet

Billet

Roughing and
forming passes

Finishing pass

Several intermediate passes have
been omitted for clarity

1-1 *Steel from a smelter batch to be heated as an ingot.* American
Iron and Steel Institute

The steel-making industry and rolling mills usually adhere to American Society for the Testing of Materials (ASTM) specifications. I wish to explain a common mistake made by welders and fabricators. The carbon contents of steels do provide absolute categories. Steels

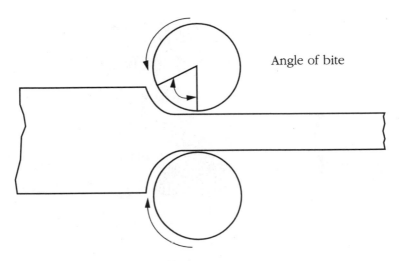

Angle of bite

1-2 *Rolls that flatten steel.*

containing 0.05 to 0.30 percent carbon are low carbon (mild steel), the structural grade A-53 now used for most mid-structural shapes. These steels are said by most experts to be able to withstand heating and cooling without it affecting the grain structure. The problem here is that steels with 0.20 percent carbon are capable of being tempered and, upon quick-quenching from 1700 degrees to below 200 degrees in two seconds, can possibly produce a *martensitic* grain structure that is extremely hard but also brittle. To further complicate the matter, many of the filler metals used in welding processes contain alloys and carbon contents well beyond the stated level.

Medium carbon steel has a carbon content of 0.30 to 0.45 percent. This medium carbon steel requires more care as welding and fabrication procedures are followed. Guidelines are given for both medium carbon and high carbon steels in Chapter 5.

Crystalline structure

The steel we use is not really the perfect, solid, continuous surface we see. The microscope and allied equipment are the tools of metallurgists, who are responsible for many of the advances in steelmaking. Under the metallurgists' microscope, we find that the surface of the steel is broken into many-sided shapes. Figures 1-4 and 1-5 are photographs taken through a microscope of mild steel used every day in shops throughout the country.

Now that we have seen the actual crystal structure, we should go one step further. The atoms that make up these crystals begin to set up

**Typical passes for 6" beams
as rolled on structural mills**

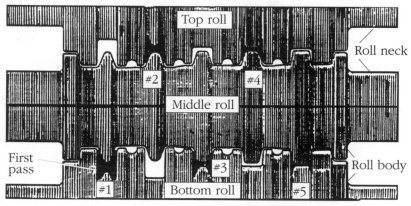

Roughers

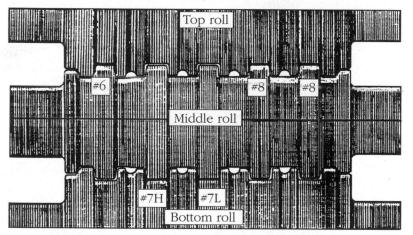

Intermediates

Finishers

1-3 *Rolls that produce diverse steel shapes.* American Iron and Steel Institute

within the crystal-forming pattern as soon as metal is no longer fluid.
These tiny particles are attracted to like particles and, in grouping
together, form the grains or crystals in steel. It is believed that the bor-
der atoms may be a little tougher or stronger than those in the center

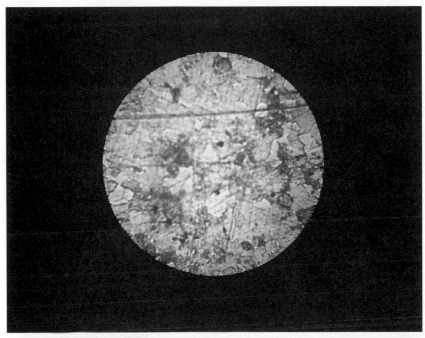

1-4 *View of the steel crystal structure in mild steel magnified 100 times.*

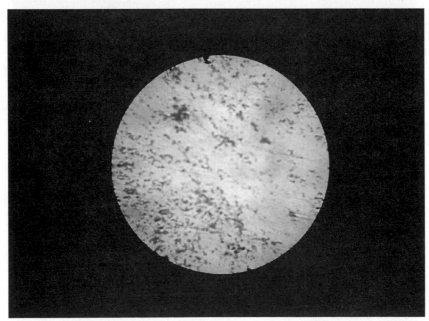

1-5 *This microscopic view of ½-inch (12.7-mm) steel from Fig. 1-4 is magnified 30 times.*

of each crystal. To the fabricator, the most important aspect of this property is in the bending or forming of steels. When bending within the proscribed limits occurs, a path between rows of atoms is followed and the crystals are not ruptured. Because of this property, it is even possible to change the shape of the crystals. The paths along which the changes occur are called *slip planes*.

Plasticity

Time and temperature also enter into the plasticity of steels. Temperature changes may result not only in the change of the size of crystals, but in the atom formations as well. Any good text on physical metallurgy is excellent background reading for those entering the field of steel fabrication.

Another factor of extreme importance to the fabricator is the *elastic limit* of steel. The elastic limit simply means the amount, measured in pounds per square inch (psi) of material, that steel will stretch and still return to an unstrained condition. Just beyond this point is the *yield point*, at which the material will continue to pull apart even though no more pressure is applied. Figure 1-6 shows a sample test strain curve. Figure 1-7 shows the machine that made the test shown in Fig. 1-6.

To better understand the workings of such equipment, we can use a very practical example. Many of the welder qualifications tests toward welder certification require the tensile testing of welds made by the welder taking the test. In reality, the steel to be welded and the filler metals are also tested at that same time. The bottom-heavy steel platform contains a serrated jaw device. This block is driven by the motor at its right end. The motor is geared down to very low revolutions per minute (rpm). It travels up or down on two large-diameter threaded columns. They are the two dark columns and arc centered between the four support shafts shown in Fig. 1-7. The top block also contains a matching jaw device. The qualification coupon shown in Fig. 1-8 has been tested to destruction.

The hand cranks (handles) merely hold the specimens in the jaws until they are clamped by the downward travel of the bottom block. The more pressure that is applied, the harder the bite of the jaws. The thousands of pounds of pressure are clearly registered by the pointer centered in large white-faced dial. At the same time, a record is made on the graph paper shown to the right of the dial. As the strain becomes more than the steel can stand, it reaches the yield point. If you were to watch the dial pointer, you would see a momentary stop and then a climb to the ultimate tensile strength and break point. These points will be clearly shown on the graph. If you look closely at that

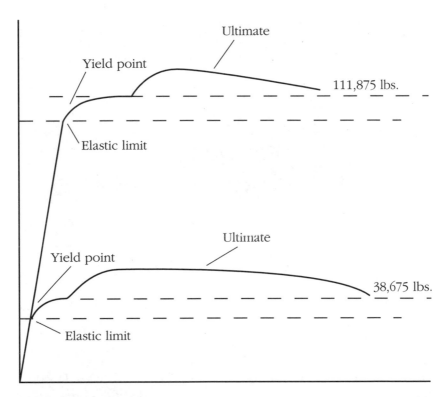

1-6 *The typical yield point and tensile strength of steel.*

1-7 *A machine for testing the yield point and ultimate tensile strength of steel.* Lane Community College

1-8 *Weld specimens tested to destruction.*

bottom specimen in Fig. 1-8, you will see a necked (area reduced in thickness and width) length of steel on either side of the break. This is exactly what the inspector wants to see. The coupon did not break in the weld. The welded area is in the center. The reduced section is well to the left of the break. Both the strength of filler metal and the quality of weld were better than the original steel. Now, if the parent metal (original steel) had the tensile strength listed for its quality number, the entire test was a success. The specimen at the top of Fig. 1-8 is another matter. The reduced section has a smooth finish. Very little necking is visible. This should lead you to suspect that this test was of the steel. No weld outlines are visible. Since the steel snapped, the yield point and ultimate tensile points were close together. This may have been an alloy or medium or high carbon steel.

To better understand the stress/strain relationship, consider what would happen if you chinned yourself on a steel bar. This exercise is well within your tolerance limits. While you were in that position, if a 200-pound weight was suddenly attached to your ankles, the stress and the damage to your arm and shoulder muscles would be strain. It could result in shoulder separation and tearing. Keep in mind that strain is the result of stress.

The next factor of prime importance to the fabricator-welder would be *elongation*. This term denotes the amount of permanent plastic extension before the elastic limit is reached. It is usually shown as the percentage of a given length. Welders recognize a percentage factor as found in 2 inches of deposited weld metal. Of the utmost importance to the welder-fabricator is that all elongation factors decrease as temperature decreases.

One very important tool of the metallurgist is shown in Fig. 1-9. It is a hardness tester. For our purposes, the relative hardness of structural steel and those beyond the common usage are considered. Just below the white-faced dial at the top center of the tester is a short metal tube. This tube accommodates a smalltool bit, which is about ⅝ inch (15.8750 mm) in length with a diamond point. The bottom pedestal with the black steel cap can be raised or lowered to accommodate the steel to be tested. The steel is brought into contact with the diamond point of the indenter. The turning of one or more of the three spoke arms do the job. The instructions for load and operation are clearly printed just below the dial. The tester operates on the depth of penetration of the indenter. The resistance is measured and recorded on the dial. Of course, the harder the steel, the more problems for the fabricator and welder. The identification and solution to many of these problems are discussed in subsequent chapters. One of the projects included in this text shows you the use of simple tools to identify relative hardness. It also details the making of a small inexpensive tool to include in your tool box or kit.

1-9 *A metallurgy tool for testing the relative hardness of steel.* Lane Community College

Always preheat any steel that is more than 1½ inches thick before welding. If the on-site temperature is 32 degrees F or less, preheat should be applied for steels down to ¾ inch thick. Failure to follow these guidelines results in almost all of the underbead cracking. It is a metallurgical problem. That part of the parent metal is already cold and does not change rapidly enough to accommodate the 10,000 degrees F heat of arc or the 2750 degrees F melting point of the filler metal and the steel it fuses to. Since underbead designates that the bead is then covered with more passes, you will not see the cracks until testing is done. Repair requires removal of all weld metal and some preparation of the parent metal before restarting the welding process. Many small fabrication companies do not understand that metal preparation and welding processes are not the same on January 3 and July 3. Expansion and contraction forces are the problem, and preheating is the answer.

If oxy-fuel equipment is not available at a construction site, use small-diameter filler metals at increased amp rates. If ⅛-inch coated electrodes are used at ³⁄₁₆ amp and volt settings, the rods will smoke and turn limp, but at least half of the electrode will produce a bead. Grind the surface smooth and proceed. If only the welding machine is available, clamp both the ground and gun cable to the steel. Do not use amp rates beyond the duty cycle of the machine, which may be as low as 30 percent of top setting for that machine. It will take some time to reach a suitable temperature, but it is better than doing it over. This process can also be used for annealing; when a slow cooling rate is required after steels are joined and the rate is too rapid, the change will result in dissimilar grain structure problems.

A weld deposit with a 16-percent elongation factor at 72 degrees F may have a 2-percent factor at 0 degrees F, making it about as brittle as cast iron. Those persons living in areas of extreme cold or shipping to those areas should use caution in choosing welding materials. Every good supply house will have electrode and wire specifications available for all the products it handles. Table 1-1 shows some of the mechanical properties of the materials you are using.

Now let's take a look at what heat does to steel. I imagine several complete books could be written on this factor alone, but for our purposes, a few highlights will be of real assistance to the fabricator. Some reasons why steel becomes more ductile with heat are quite apparent in Fig. 1-10. Not only does the grain enlarge, but the planes become more pronounced and even seem to align more readily. The steel changes of forming and bending will follow the slip planes without the resistance of forcing further changes in crystal shape. It should be remembered that the smaller the grain, the harder the steel.

Table 1-1. Some factors to consider when choosing electrodes

AWS class	Tensile strength in P.S.I.	Yield point strength in P.S.I.	Elongation in 2″ (50.8 M.M.) by percentage
E-6010	(27.5 Metric Tons) 62000	(22.6 M.T.) 50000	22%
E-6011	(27.5 M.T.) 62000	(22.6 M.T.) 50000	22%
E-6012	(30.1 M.T.) 67000	(25 M.T.) 55000	17%
E-6013	(30.1 M.T.) 67000	(25 M.T.) 55000	17%
E-6014	(27.5 M.T.) 6200	(22.6 M.T.) 50000	17%
E-6020	(27.5 M.T.) 6200	(22.6 M.T.) 50000	25%
E-6027	(27.5 M.T.) 6200	(22.6 M.T.) 50000	25%
E-7010	(33 M.T.) (37.5 M.T.) 72000-81000	(28.5 M.T.) (32.0 M.T.) 63,000-71000	24%-32%
E-7014	(33 M.T.) 72000	(27.2 M.T.) 60000	17%
E-7015	(33 M.T.) 72000	(27.2 M.T.) 60000	22%
E-7016	(33 M.T.) 72000	(27.2 M.T.) 60000	22%
E-7018	(33 M.T.) 72000	(27.2 M.T.) 60000	22%
E-7024	(33 M.T.) 72000	(27.2 M.T.) 60000	17%
E-7028	(33 M.T.) 72000	(27.2 M.T.) 60000	22%

Extreme care must be taken with some steels. Manganese steel, on application of heat above 500 degrees for long periods of time, becomes soft and fibrous, losing almost all its toughness and strength. Many of the low-alloy steels, which derive their strength from the quick quench, must be cold-formed and should never be subjected to heat above 1200 degrees. These steels are part of the structural family and are mentioned in another part of this chapter.

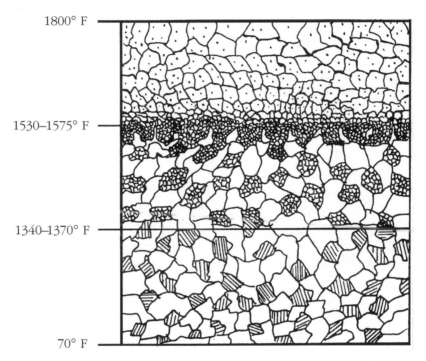

1-10 *Structural changes of A-36 mild steel due to heating.*

The annealing and normalizing processes are vastly underused because of cost factors. Almost all small fabrication shops lack the expensive, large ovens needed to work in these fields. Post-heating and preheating of weld areas would relieve grain stress and residual stress. Anyone dealing with high carbon and alloy steels will be more aware of the problems because they have seen the effects firsthand. When the steel becomes molten during the welding process, the grain sizes have not had sufficient time to become alike in structure. The carbon has a tendency to precipitate outwardly, and hydrogen is introduced into the weld puddle. As the weld area cools, contracting grain structure on an uneven scale and radiating out from the fusion zone, it sets up additional strain patterns. All welding does follow the application of metallurgy, both theory and practice.

The word *theory* is poor usage. We cannot use an electron microscope to examine steel, because it is opaque in any thickness, unlike animal or leaf tissue that light passes through easily, so the things we know cannot be easily proven. We can cross-section a weld in steel and, with proper preparation and magnification, see the changes in crystal formation caused by the heat, the filler metal, and the distance from the weld center at which the steel has been affected. Note that in Fig. 1-10, cooling to or not reaching 200 degrees F (93 degrees Celsius, or C) does not affect the grain structure. The steel was pulled (warped) by the shrinkage in the heat-affected zone (HAZ) as it cooled. Actually parent metal would be similarly affected. In Chapter 8 you see how proper positioning or restraining of the parent metal minimizes this factor. A second weld made on the other side of the joint won't bring the parent metal back to perpendicular, because once shrinkage (grain change) has occurred, it will not be affected to nearly the same amount a second time. It is also apparent that the first weld plus the thickness of the parent metal is much greater that the second side weld deposit. Preheating, plus the use of ductile (good elongation), compatible, high-tensile, but low-hydrogen electrodes and normalizing will prevent most weld fracture patterns.

Referring back to Fig. 1-10 may help you to understand the ease with which mild steels are hot-formed. At about 1350 degrees F, the critical temperature of the steel is reached. A full but dull red color is attained. At this point, the atomic structure of the steel is changed from body to face-centered cube lattice. Of more importance to the fabricator is the fact that the steel is now nonmagnetic, and the phase of the steel is spheroidizing. Just as if tiny ball bearings were formed to roll one crystal over another, so now the steel seeks to follow the law of gravity.

The linear coefficient of expansion in steel is also relatively simple to explain. For each degree of heat, each square inch of steel affected increases in size 0.000006 inch. A piece of flat mild steel allowed to lay in the sun until its temperature increased from 50 to 110 degrees F would increase in size by 60 × 0.000006 inch for each square inch of area. A steel plate 10 feet by 20 feet increases in size according to this formula:

10 feet × 20 feet × 144 inches/square foot × 60 degrees change
× 0.000006 inch = 10.37 square inches

Cold-quality steel is a higher-strength steel with some cold-bending limitations. This type of steel is for shipbuilding, but it is not pressure-vessel quality.

Forge-quality steel is *killed steel*, which is generally free of surface defects and must pass rigid, bend, and tensile tests. Flange-quality is pressure-vessel quality, except when it exposed to heat. Firebox-quality steel may be exposed to mechanical stress and heat in pressure vessels. Marine-quality steel can be used in combustion chambers and pressure vessels in marine boilers. All of these steels are made to ASTM specifications.

Grades of steel may follow military specifications and numbering systems. Almost without exception the military specs call for higher tensile strengths than comparable standard structural grade steels (Table 1-2). Also notice that other specification bodies are included.

Grade A steels are of lower tensile strength than grade B steels. The minimum ultimate tensile strength for grade A is 45,000 psi. For grade B (used in most American Society for Mechanical Engineers pressure-vessel weld tests), the minimum allowable strength is 58,000 psi.

Since about 90 percent of all steel used in the United States is mild steel or low carbon steel, it is very important that you realize what welding does to these steels. You know that steel undergoes change as it is heated. The ability of steel to exist in two or more crystal structures at different temperatures is a real transformation. At low temperatures, steel has a body-centered atomic structure and is called alpha phase. At high temperatures, it changes to a face-centered atomic structure and is called delta phase.

Fusion welding is almost a resmelting operation. The standard stick-welding process produces a heat of approximately 10,000 degrees F, or 6000 degrees C. The welding range of mild steel for this process is 2400 degrees F, to the melting point of approximately 2850 degrees F.

Referring back to Fig. 1-10, we see the dark grains, pearlite and ferrite (iron), at low temperatures. Above the 1340–1370-degree line we see ferrite and austinite. Austinite is the friend of welders; it embodies ductility and strength without brittleness. This property continues on until the liquid delta point of approximately 2850 degrees. Very little detrimental change occurs in the process. Some susceptibility to oxidation occurs. Leave two pieces of steel out overnight, one with beads on it, the other without, and the welded specimen will show a marked increase in rust formation (oxidation). Also, some material may go beyond the liquid state and become gas, or gas from the atmosphere may be trapped in the cooling metal structure, a state called *porosity*. Remember, water is H_2O: two parts hydrogen gas to one part oxygen gas. The air we take into our bodies is also gas: about 70-percent nitrogen, approximately 22-percent oxygen and the remainder mostly hydrogen. As the air around us becomes saturated with moisture, humidity

Table 1-2. Specifications and tests

Specifications

The majority of the carbon steel plates and structural sections are produced to meet specific standard specifications. These specifications are usually prepared by one of the specification writing bodies listed opposite.

These specifications are referenced by regulatory bodies who are concerned with the public's safety when using buildings, structures, vehicles, and vessels. The specifications are constantly being updated by their originating bodies. Consequently, these bodies should be consulted regarding available standard specifications.

Abbreviation	Full Title
AAR	Association of American Railroads
ABS	American Bureau of Shipping Rules
AREA	American Railway Engineering Association
ASME	American Society of Mechanical Engineers
ASTM	American Society for Testing and Materials
DOD	United States Government— Department of Defense
USCG	United States Coast Guard

Scope of Available Standard Specifications

Designation	Specification Title	Specifications Writing Bodies
Ship Steels	Structural Steel for Hulls Structural Steel for Ships	ABS, ASTM
Structural Steels	Carbon Steel Plates, Low and Intermediate Tensile Strength, Structural Quality	ASTM, ASME
	Structural Steel for Locomotives and Cars	ASTM, ASME AAR
	Steel, Structural Shapes, Plates and Bars	AAR, ASTM
	Structural Steel for Welded Structures	ASTM
	Steel for Bridges and Buildings	ASTM
	Structural Steel-Iron and Steel Structures	AREA
Carbon-Silicon Steels	Low and Intermediate Tensile Strength Carbon-Silicon-Steel Plates for Machine Parts and General Construction	ASTM
Ordinary Flange and Firebox Steels	Low and Intermediate Tensile Strength Carbon-Steel Plates of Flange and Firebox Qualities	ASTM, ASME
Flange and Firebox Carbon-Silicon Steels	Carbon-Silicon-Steel Plates of Intermediate Tensile Ranges for Fusion-Welded Boilers and Other Pressure Vessels	ASTM, ASME
	High Tensile Strength Carbon-Silicon-Steel Plates for Boilers and Other Pressure Vessels	ASTM, ASME
	High Tensile Strength Carbon-Manganese-Silicon Steel for Boilers and Other Pressure Vessels	ASTM, ASME
	High Tensile Strength Carbon-Manganese-Steel Plates for Unfired Pressure Vessels	ASTM, ASME
Marine Steels	Marine Boiler Steel Plates	USCG

American Iron and Steel Institute

increases, which means more gas may be introduced into the weld puddle directly from the atmosphere, resulting in porosity.

Figure 1-11 shows the pull of the welds as they cool. Mill practices are neither mistakes nor learning processes. They are rigid rules and tend toward the most efficient and best way to produce quality structural shapes.

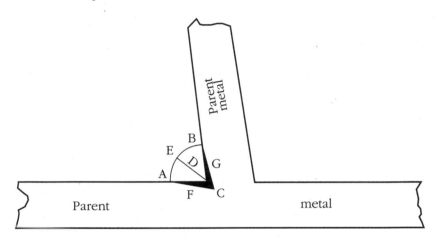

1-11 *The pull of the weld on steel.*

As steels cool from the smelting process, several types of steel are formed. If no trapped gases cause raising of the "crust" on an ingot and it lies quietly, it is *killed* steel. *Semi-killed* has a secondary pipe or cavity and gases usually evolve more slowly. *Rimmed steels* have marked differences in chemical composition. The outer rim is lower in carbon, phosphorous, and sulfur than the rest of the ingot. The core has more of these elements. The oxidizers are virtually eliminated from rimmed steels (aluminum, silicon, and titanium).

Commonly specified elements are shown in Tables 1-3 and 1-4. One further note should be added on sulfur steels. Weldability decreases in cases where these steels were involved in fabrication of nuclear reactor shells in combination with copper. The life of the shell was cut almost in half. The alloys, their contributing factors and uses, are covered in Chapter 5.

In mill practice, sheared plate will actually be above the theoretical weight of a section. Most plate has a given weight factor of 40.8 pounds per square foot per inch of thickness. Do not measure thickness within ⅜ inch of any edge, because shearing and rolling may affect such edges.

Table 1-3. Ladle chemical ranges and limits carbon steels (plates and structural sections)

Chemical Ranges and Limits, Per Cent

(1) Element	(2) When Maximum of Specified Element is	(3) Range	(4) Lowest Max.*
CARBON			0.12
(Note 1)	To 0.15, incl.	0.05	
	Over 0.15 to 0.30, incl.	0.06	
	Over 0.30 to 0.40, incl.	0.07	
	Over 0.40 to 0.60, incl.	0.08	
	Over 0.60 to 0.80, incl.	0.11	
	Over 0.80	0.14	
MANGANESE			0.40
	To 0.50, incl.	0.20	
	Over 0.50 to 1.15, incl.	0.30	
	Over 1.15 to 1.65, incl.	0.35	
PHOSPHORUS			0.040**
	To 0.08, incl.	0.03	
	Over 0.08 to 0.15, incl.	0.05	
SULPHUR			0.050**
	To 0.08, incl.	0.03	
	Over 0.08 to 0.15, incl.	0.05	
	Over 0.15 to 0.23, incl.	0.07	
	Over 0.23 to 0.33, incl.	0.10	
SILICON			0.10
	To 0.15, incl.	0.08	
	Over 0.15 to 0.30, incl.	0.15	
	Over 0.30 to 0.60, incl.	0.30	
COPPER	When copper is required, 0.20 per cent minimum is commonly specified.		

* Lower maxima can be produced.

** Lower phosphorus and sulphur maxima are furnished for some qualities as noted in Section 5.

Note 1. Carbon: When the maximum manganese limit exceeds 1.00 per cent, add 0.01 to the carbon range shown in Column 3.

American Iron and Steel Institute

Table 1-4. Commonly specified elements

It is the purpose in the next few paragraphs to outline briefly the effects of the elements carbon, manganese, phosphorus, sulphur, silicon and copper on manufacturing practices and the steel's properties. The effects of a single element on either practice or properties are influenced by the effects of other elements. These interrelations, frequently of a complex nature, must be considered when evaluating a change in specified composition. However, to simplify this presentation, the various elements will be discussed individually. The scope of this discussion will permit only suggestions of the modifying effects of other elements, or of steelmaking practices, on the effects of the element under consideration.

The first four elements briefly discussed in the following paragraphs are those most generally specified in carbon steel.

Carbon   The surface quality becomes impaired as the carbon content increases in rimmed steels. By contrast, killed steels have poorer surface in the lower carbon grades. Carbon segregates within the ingot and, because of its major effect on properties, carbon segregation is frequently of more significance and importance than the segregation of other elements.

Carbon is the principal hardening element in steel and as carbon increases the hardness of steel increases. Tensile strength increases and ductility and weldability decrease with increasing carbon content.

Manganese   has a lesser tendency than carbon to segregate within the ingot. It is beneficial to surface quality in all carbon ranges, and is particularly so in high sulphur steels.

Manganese contributes to strength and hardness but to a lesser degree than carbon. The amount of increase in these properties is dependent upon the carbon content, i.e., higher carbon steels are affected more by manganese than lower carbon steels. Increasing the manganese content decreases ductility and weldability, but to a lesser extent than carbon.

Phosphorus has a tendency to segregate within the ingot being exceeded in this respect usually by sulphur and carbon.

Generally, increased phosphorus results in increased strength and hardness and decreased ductility. This is particularly true in higher carbon steels that are quenched and tempered. Phosphorus improves resistance to atmospheric corrosion.

Sulphur has a greater tendency to segregate within the ingot than any of the common elements. It is detrimental to surface quality particularly in the lower carbon and lower manganese steels. Resulphurized steels have a higher sulphide inclusion frequency than equivalent grades of non-resulphurized steels.

Generally, sulphur decreases ductility, toughness and weldability. Sulphur is beneficial to machinability, and the improvement in this characteristic is the only reason for adding sulphur to steel.

Silicon is one of the principal deoxidizers used in steelmaking and therefore the amount of silicon present is related to the type of steel. Silicon is somewhat less effective than manganese in increasing strength and hardness. It has only a slight tendency to segregate within the ingot.

Copper has a moderate tendency to segregate within the ingot. Since copper is not removed by any of the conventional steelmaking processes it is becoming increasingly difficult to maintain low copper maxima. Copper is detrimental to surface quality and exaggerates surface defects inherent in high sulphur steels.

Copper in appreciable amounts is detrimental to hot working operations. In the small amounts used in carbon steels, copper has no significant effect on mechanical properties. When copper is present in low carbon steel in excess of 0.15 per cent it has a beneficial effect on atmospheric corrosion resistance.

Since sheared plate does not guarantee squaring of edges, the edges should be checked. They should not be out of tolerance more than $\frac{1}{4}$ inch in width plus the camber tolerance, plus $\frac{1}{2}$ the nominal thickness of the plate. For the resquaring of carbon steel plates by gas cutting, Table 1-5 will be of help. Although very little attention is given to preheating the plate, it is apparent from the data in Table 1-5 that care should be used in dealing with thicker and higher-carbon plates.

Surface flaws may be chipped out and repaired by welding, but not more than 2 percent of the surface may be affected. Also, after chipping is complete, the thickness must not be reduced by more than 20 percent. Further, welding must be done by qualified persons. No undercut or overlap is allowed, and surfaces shall be $\frac{1}{16}$ inch above surface levels before grinding. All work will be inspected by experienced mill inspectors.

In accordance with American Society of Mechanical Engineers (ASME) Boiler Code Committee and various ASTM specifications, all plate of flange, firebox, and other specific quality are subject to slab or ingot identification for reference to test data under ¼ inch by stencil or ¼ inch by die stamp. The rolling mills interchange the terms shape and section.

The same rules apply for repair of surface defects in shapes or sections as for plate, with the exception of depth of repair, which may not exceed 30 percent of depth and not over ½ inch (12.7 mm) in any case. The weld will be of low-hydrogen quality filler materials and conform to all above specifications.

Table 1-6 gives specifications and test data from the American Iron and Steel Institute (AISI). Sections may be tested to almost any code specification if required by the consumer. These include nondestructive methods, such as magnetic particle, X- ray, and ultrasonic inspection.

Table 1-5. Customary location of test specimens for ASTM and similar steel plate specifications

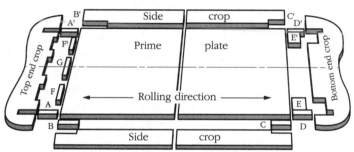

	Tensile	Bend	Homo-geneity	Check Analysis	Impact	Non-metallic
Structural Quality	C, C', D or D'	C, C', D or D'		C, C', D or D'		
Flange Quality	C, C', D or D'	G*		C, C', D or D'		
Firebox Quality	A, A', B or B', & C, C', D or D'	G*	A, A', B or B'	A, A', B or B', C, C', D or D'		
Special tests					C, C', D, or D', F* or F'*	E or E'

* Transverse specimens—all other test specimens are longitudinal. Other special tests are taken from any location as established between consumer and producer. Drillings for check analysis can be obtained from broken tensile test specimens.

Sampling for Tension Test
and Bend Tests

The number of tension test specimens and bend test specimens obtained for the several plate qualities, as shown in ASTM specifications, is given below.

Structural Quality Plates and Structural Quality Sections.
Specimens for two longitudal tension tests and two longitudal bend tests are obtained from each heat, unless the finished steel from a heat is less than 30 tons in which case one tension specimen and one bend specimen are obtained. However, if plates or structural sections from one heat differ $\frac{3}{16}$ in. or more in thickness, one longitudinal tension specimen and one longitudinal bend specimen are obtained from both the thickest and thinnest product rolled regardless of the weight represented. Tension and bend test specimens are selected from the webs of beam sections, channels and zee sections; from the legs of angles and bulb angles and stems of rolled tees.

Qualities Requiring Plate-As-Rolled. *
When tests are required from each plate-as-rolled a longitudinal tension test specimen, for determining the acceptable minimum tensile strength, yield point and ductility, is taken at the bottom corner of the plate. A transverse bend test specimen is taken from the middle of the top end of the plate-as-rolled. When required a longitudinal tension test specimen is taken at a top corner to determine tensile strength. Chemical analysis may be determined from a broken tension test specimen.

Longitudinal Test,
unless specifically defined otherwise, when it refers to a test specimen, means that the length wise axis of the specimen is in the direction of the rolled extension of the shape or greater rolled extension of the plate. This means that the stress applied in testing tension test specimens is in the direction of the greater rolled extension of the plate and the fold of the bend test specimen runs crosswise (transversely) to the greater rolled extension.

* Plate-as-rolled refers to the original total product rolled from a slab or directly from an ingot.

Table 1-5. *(continued)*

	There is an important distinction between the terms "longitudinal bend" as defined above with respect to test specimens and the term "longitudinal bend" with respect to forming or bending in fabrication. In the latter case, the fold of the bend runs parallel (longitudinally) to the greater rolled extension; that is, "with the grain."
Transverse Test,	unless specifically defined otherwise, when it refers to a test specimen, means that the lengthwise axis of the specimen is at right angles to the direction of the greater rolled extension of the plate from which it is cut. This means that the stress applied in testing tension test specimens is at right angles (transverse) to the greater rolled extension of the plate and the fold of the bend test specimen runs parallel (longitudinally) to the greater rolled extension.
	There is in important distinction between the term "transverse bend" as defined above with respect to test specimens cut from plates and the term "transverse bend" with respect to forming or bending of the plates in fabrication. In the latter case the fold of the bend runs at right angles (transversely) to the greater rolled extension, that is, "across the grain."
Number of Tests.	For the number of tests applicable to any quality refer to Quality Descriptions, pages 27 to 29.
	test specimens cut from plates and the term "transverse bend" with respect to forming or bending of the plates in fabrication. In the latter case the fold of the bend runs at right angles (transversely) to the greater rolled extension, that is, "across the grain."
Number of Tests.	For the number of tests applicable to any quality refer to Quality Descriptions.

Table 1-6. Tests

Carbon steel plates and structural sections are produced to one of the following:

(a)   Mechanical test requirements and no chemical limits on any element.

(b)   Mechanical test requirements and chemical limits on one or more elements, when such requirements are technologically compatible.

(c)   Chemical limits on one or more elements and no mechanical
        test requirements.

Mechanical Test Requirements

Mechanical testing of carbon steel plates and structural sections includes
tensile tests and bend tests. Those and some other mechanical tests are
described in ASTM Designation A 370.
Some significant aspects of mechanical testing of carbon steel plates and
structural sections are given below.

(1)   Variations in chemical composition, in manufacturing processes and
        in product form and dimensions result in variations in mechanical
        properties. Those variations suggest the following tensile strength
        ranges.
        10,000 psi for minima to 56,000 excl.
            (example, 55,000 to 65,000 psi)
        12,000 psi for minima 56,000 to 67,000 psi, excl.
            (example, 65,000 to 77,000 psi)
        15,000 psi for minima 67,000 to 80,000 psi, incl.
            (example, 70,000 to 85,000 psi).

(2)   The ratio of minimum yield point to minimum tensile strength for
        carbon steel is inherently established by chemical composition and
        treatment. That ratio is recognized in the yield point and tensile
        strength limits of applicable ASTM specifications.

(3)   Elongation and bend test requirements, consistent with mechanical
        properties of carbon steel, have been established in ASTM
        specifications.

(4)   ASTM specifications contain requirements for the selection of test
        specimens and number of tests, and the selection and number are the
        results of experience.

Additional Mechanical Tests

Some of the additional mechanical tests that can be made on carbon
steel plates and structural sections include the following, which are
described in ASTM Designation A 370.

(a)   Charpy impact test.

(b)   Measurement of elastic properties by an extensometer, or by a
        method that requires plotting a stress-strain diagram, or its
        equivalent, as in the cases of elastic limit, proportional limit, yield
        strength by the offset method or extension under load.

(c)   Brinell hardness test.

Project 1: Basic metallurgy tests

For common testing of steel, you need a few simple tools: a standard file capable of an edge cut (even a triangular file will do); a small magnet, either horseshoe or bar shape; a center punch; and a scribe. You will also need an oxy- fuel torch and, in one case, a welding machine and some stick electrodes.

You can make one tool for testing the relative hardness of steel. You need a steel ball bearing; any diameter from ¼ inch (6.35mm) through ½ inch (12.7 mm) is acceptable. Any machine shop or fastener supply house will have many different sizes. You might be able to get one free of charge. You also need a piece of *clear* plastic tube 6 inches (152.4 mm) in length. The inside diameter of the tube should match the size of your ball bearing. All building supply stores will have the tubing. Set the 6-inch (152.4-mm) tube upright on a piece of low carbon mild steel (A-36). Drop the ball down the tube. Note how high the bearing bounces. Mark the height of the bounce on the tube with your scribe. Test this device on stock of increasing hardness (carbon content). The harder the steel, the higher the bounce. Mark the tube for each type of steel.

While some steels can be cut with a power shear or an iron worker, some cannot. Even low-alloy, high-tensile-strength steels mentioned in Chapter 3 can damage shear blades and iron-worker punches and blades. Check with a supervisor before the operation of such equipment. In general, if a shear cuts ½-inch (12.7-mm) A-36 steel, it can safely handle ¼-inch (6.35-mm) low alloy. Check before cutting any steel other than low carbon. You cannot drill many of the *hardened* low-alloy steels. In Chapter 3, one steel has a Brinell hardness of 321, which is approximately 34 on the Rockwell C scale.

Some of the steels you might test are much harder than standard cutting blades. This hardness tester also tells you which steels will present welding problems. Since fabrication is our real concern, you can usually assume that the shapes (beams, channels, angles, etc.) are mild or low carbon steel. Some low-alloy, high-tensile-strength steel shapes are now sold. They should be clearly marked. Any scrap or remnant steels not marked should be tested. This is especially true of plate or bar stock.

In using the file to check hardness, always use the edge on a plate or bar edge corner. Structural steel should show a pronounced bright groove. Using the same hard stroke of the file edge will show increasingly *less* depth of groove as the hardness of the steel increases. Of course, the harder the steel, the more problems you will encounter in fabricating and welding. Abrasive-resistant (AR) plate and those steels having 0.6 percent or more carbon will tear teeth from the file. The welding of some of these steels is detailed in Chapter 10.

The magnet separates the nonferrous metals from steel instantly. The trouble here is that manganese (Hadfield grade, approximately 18 percent) is also nonmagnetic. Many of the stainless steels are also nonmagnetic.

The torch has many uses as a testing tool. To separate the stain-lesses from manganese, simply try a cut. The oxy-fuel torch will cut manganese, but it will not cut stainless. It will cut cast steel but not cast iron. It will melt aluminum, but it will burn magnesium filings with a sparkling white heat. *Do not* set a large piece of this metal on fire. Magnesium is used in incendiary devices and flares. It burns well under water, obtaining oxygen from the water itself. Dry sand will put it out. The sand melts and forms a glass cover that denies oxygen to the flame. If you suspect cast iron rather than steel (both have a rough-appearing surface), strike an arc on the surface. Your file will not even scratch that area if it is cast iron. You can still file cast steel. Cast iron without the arc strike is soft in relation to cast steel and files with a deeper groove than A-36.

If you are good at measured hammer blows, the depth of pene-tration from the center punch is another way of checking hardness factors. If you remember the facts presented here, you should have few failures. Keep the tools ready for use and the torch and welders nearby. If this is a class project, show the tools and the results of your findings to your instructor.

Project 2: Cambering

This project is truly a study of the effects of heat on metal. While the use of hand torches is not detailed until Chapter 6, your instructor or mentor can show you the safe setup and operation of the equipment. The use of heating tips for the combination torch requires more gas pressure than a standard cutting operation. The recommended gas pressures for a metal thickness of ½ inch to ¾ inch (12.7 mm to 19.05 mm) is 10 to 14 psi (0.7 to 1.0 kilograms per square centimeter) of oxygen and 6 to 9 psi (0.41 to 0.635 kilograms per square centimeter) of acetylene for a #6 heating tip or nozzle. The first area to be heated is the exact center of the beam. Since we do not know the size beam available to you, we can give you the correct data in general. Measure the width of the top flange. Heat an area equal to the width. Heat the entireV-shaped area as shown in Fig. 1-12, including the bottom flange area, to relax the rigid holding strength of the beam section (shape) at this point. When the red color is uniform for the area you are heating, the heating is sufficient. This color means you have reached or passed the crystal phase of change. One theory holds that the grains have become round (spheroid) and the steel is nonmagnetic. The grains are free to move and follow the law of gravity.

If you now wish to enhance the change, cool the area with rags soaked in cold water. The area changes back to crystal formation and also shrinks the spaces between the crystals and perhaps the grains themselves. If you stretch a line from end to end of the beam, you will see a downward bow. Now continue to heat the areas of beam as shown. When the cooling process is complete, again measure the downward curve. Turn the beam over. It will have a nice, even, pronounced arch. This arch is also called camber.

If you need to straighten the beam to its original shape, heat the same size area on either side of the previously shrunk metal. Once the steel has set, it will not respond in the same manner. You cannot simply reverse the process in the same area as before. The dotted lines show the correct way to heat and remove the arch.

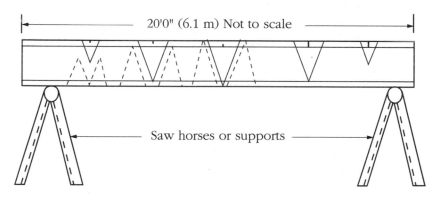

1-12 *The sketch view of the points of heat application to camber a beam.*

Questions for study

1. Name three forms of steel supplied to the rolling mill by the smelter.
2. Which way do crystals lie after being formed by rolling?
3. Is steel really solid? If not, what part is not solid?
4. Are cracks in steel always a straight line? If not, why not?
5. What is the approximate carbon content range of mild steel?
6. At what point does the carbon content in steel affect its hardness?

7. What is a slip plane?
8. Do crystals in steel change shape or size? If so, what is the main factor in causing the change?
9. How can we make the bending of mild steel easier?
10. If we raise the uniform temperature of a piece of flat mild steel 200 degrees, and the piece is one inch thick, 20 feet long, and 12 inches wide, how much will it increase in length?
11. Which grade of steel, A or B, have a higher tensile strength?
12. What is the approximate welding range of mild steel?
13. What is the melting point of low carbon steel?
14. What is oxidation?
15. What elements are found in water?
16. What does the abbreviation ASME stand for?
17. Why are so many steels listed as structural steels in Table 1-3, and what percentage of steel used in the United States is of this type?

2

Shapes and tolerances

The best way to illustrate the strength of shapes is to show that flat material is only strong in relation to thickness. If we take a dollar bill, a French franc, or Japanese yen, the paper currency has little structural strength. Place one across the open mouth of a drinking glass set in an upright position. Place a coin of almost any weight in the center of the paper currency. The weight will collapse the paper and drop into the glass. Now fold the bill in accordion pleats as shown in Fig. 2-1. This shape will support many coins.

I will list at this time most of the types of shapes in general use in the United States. The first common shape is flat mild steel. Flat mild steel is available as sheet stock, coil stock, flat bar, and universal mild plate (UMP).

Sheet stock is under $\frac{3}{16}$ inch thick, usually identified by gauge in either numerical or letter designations. Table 2-1 shows the standard gauges for wire, sheet, and plate steel. The various types of gauges overlap each other, so you must always identify the type as well as the number or letter of that particular gauge. From Table 2-1, we see that 16-gauge would be 0.0625 inch thick using the U.S. Standard. In Brown and Sharp, 16-gauge would be 0.0508 inch thick.

Coil stock is mostly for gauge-designated material and simply rolled to avoid permanent kinks or twists. Storage is also a factor. Coils are narrower than sheets and for most use do not exceed 3 feet in width (Fig. 2-2). Coil stock, especially in nonferrous metals, may be used as shim stock, only a few thousandths of an inch thick and a fraction of an inch wide. Coil stock is often used for cold-formed stamping work and punch-press work. These tools are shown in Chapter 7.

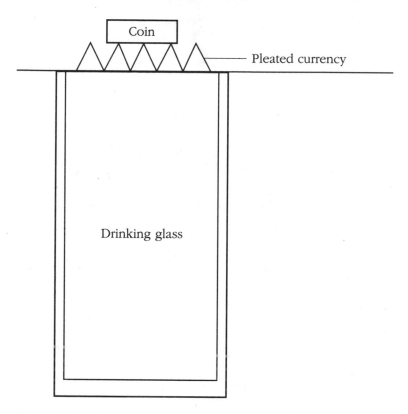

2-1 *This simple experiment demonstrates the effect of shape on strength.*

Flat bar is usually hot-rolled mild steel and, for common use, overlaps coil and sheet stock sizes. It is available from most supply houses, from ⅛ inch thick and ½ inch wide to 3 inches thick and 8 inches wide. Its identification on a blueprint is FMS (flat mild steel).

UMP is universal mild plate. Depending on the rolling mill, it usually starts at 3/16 inch thick and 9 or 10 inches to 30 or 36 inches wide. Like flat bar, it is available to 3 inches thick. It is also hot-rolled mild steel and fills the gap between flat bar and flat plate sizes. Your local steel supply warehouse has a complete listing of sizes and shapes available for immediate purchase or ordering.

In standard mild steel plate, United States Steel Corporation lists 3/16-inch plate to 20 feet wide and 60 feet long. These widths and lengths are available though 15 inches in thickness. Figure 2-3 shows some plate thicknesses and sizes stocked by a local warehouse.

To more thoroughly understand structural shapes, how they function, and the "laws" under which they operate, we must refer back to Fig. 1-2. As these pieces are rolled into various shapes, the

Table 2-1. Standard gauges of sheet or coil stock

Decimal Equivalents

Gage No.	Birmingham Wire (B.W.G.) also Stubs Iron Wire	Brown and Sharpe or American Wire (A.W.G.)	Manufacturer's Standard Mfr's. Thickness for Steel	British Imp. Standard Wire (S.W.G.)	Stand, Steel Wire Gages or Washburn and Moen	Stubs Steel Wire	U.S. Standard (old)
7-0	....	....	....	.500	.4900	....	.5000
6-0	....	.5800	....	.464	.4615	....	.4687
5-0	.500	.5165	....	.432	.4305	....	.4375
4-0	.454	.4600	....	.400	.3938	....	.4062
3-0	.425	.4096	....	.372	.3625	....	.3750
2-0	.380	.3648	....	.348	.3310	....	.3437
0	.340	.3249	....	.324	.3065	....	.3125
1	.300	.2893	....	300	.2830	.227	.2812
2	.284	.2576	....	276	.2625	.219	.2656
3	.259	.2294	.2391	252	.2437	.212	.2500
4	.238	.2043	.2242	.232	.2253	.207	.2344
5	.220	.1819	.2092	.212	.2070	.204	.2187
6	.203	.1620	.1943	.192	.1920	.201	.2031
7	.180	.1443	.1793	.176	.1770	.199	.1875
8	.165	.1285	.1644	.160	.1620	.197	.1719
9	.148	.1144	.1495	.144	.1483	.194	.1562
10	.134	.1019	.1345	.128	.1350	.191	.1406
11	.120	.0907	.1196	.116	.1205	.188	.1250
12	.109	.0808	.1046	.104	.1055	.185	.1094
13	.095	.0720	.0897	.092	.0915	.182	.0937
14	.083	.0641	.0747	.080	.0800	.180	.0781
15	.072	.0571	.0673	.072	.0720	.178	.0703
16	.065	.0508	.0598	.064	.0625	.175	.0625
17	.058	.0453	.0538	.056	.0540	.172	.0562
18	.049	.0403	.0478	.048	.0475	.168	.0500
19	.042	.0359	.0418	.040	.0410	.164	.0437
20	.035	.0320	.0359	.036	.0348	.161	.0375
21	.032	.0285	.0329	.032	.0317	.157	.0344
22	.028	.0253	.0299	.028	.0286	.155	.0312
23	.025	.0226	.0269	.024	.0258	.153	.0281
24	.022	.0201	.0239	.022	.0230	.151	.0250
25	.020	.0179	.0209	.020	.0204	.148	.0219
26	.018	.0159	.0179	.018	.0181	.146	.0187
27	.016	.0142	.0164	.016	.0173	.143	.0172
28	.014	.0126	.0149	.015	.0162	.139	.0156
29	.013	.0113	.0135	.014	.0150	.134	.0141
30	.012	.0100	.0120	.012	.0140	.127	.0125
31	.010	.0089	.0105	.012	.0132	.120	.0109
32	.009	.0080	.0097	.011	.0128	.115	.0102
33	.008	.0071	.0090	.010	.0118	.112	.0094
34	.007	.0063	.0082	.009	.0104	.110	.0086
35	.005	.0056	.0075	.008	.0095	.108	.0078
36	.004	.0050	.0067	.008	.0090	.106	.0070
37	....	.0045	.0064	.007	.0085	.103	.0066
38	....	.0040	.0060	.006	.0080	.101	.0062
39	....	.0035	....	.005	.0075	.099	....
40	....	.0031	....	.005	.0070	.097	....

Standard Gages - Wire, Sheet, Plate

2-2a *Coil or flat strip stock.*

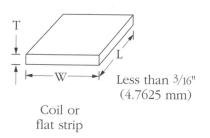

T × W × L

Less than 3/16"
(4.7625 mm)

Coil or
flat strip

T = Thickness in inches or
fractions of an inch
W = Width in inches
(usually does not
exceed 6" (132.4 mm))
L = Length in feet and/or inches

2-2b *Flat bar or flat mild steel.*

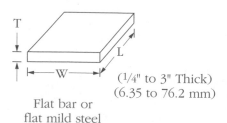

T × W × L

(1/4" to 3" Thick)
(6.35 to 76.2 mm)

Flat bar or
flat mild steel

T = Thickness in inches or
fractions of an inch
W = Width in inches
(usually does not
exceed 8" (203.2 mm))
L = Length in feet and/or inches

2-2c *Flat mild steel plate of various thicknesses.*

2-3 *Thin flat plate steel is stacked in the foreground.*

grain patterns lengthen. While the open spaces between the grains are still present, it becomes apparent that compressing the shapes from the ends may shorten the length but will not crack or crush the grains. Figures 2-4, 2-5, and 2-6 show the most commonly used shapes. As you study them, think in terms of compressing them and what the results would be.

The angles would bend out of alignment quite easily, the channels not quite so easily. The I beams and wide-flange beams would, of course, offer the greatest resistance. Figure 2-7 shows both small wide-flange beam and pipe.

Wide-flange beams may carry the symbol W- to identify them. They also are used as bearing piles because of extreme compression strength, and if flange widths are equal to the depth of section, they may be called H columns.

M and S beams were formerly listed as I beams. The first item in most steel book listings is the standard beam, and any in each depth of section thereafter is a mill beam. Of course, flange widths and web thickness are the factors to be considered.

Standard pipe is given by inside diameter (I.D.) numbers up to 12 inches, but remember, the outside diameter (O.D.) does not change on extra-strong or double extra-strong pipe. As wall thickness increases, the I.D. diminishes. This makes it possible for a single set of pipe threading dies to handle all three strengths of a pipe size. See Fig. 2-8

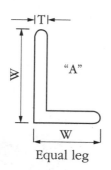

W × W × T × L

W = Width of leg
T = Thickness of leg

A 4" × 4" × 1/2" × 10'0"
   (101.6 × 101.6 × 12.7 mm × 3.1 m)

Equal leg

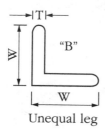

B 3" × 4" × 1/2" × 10'0"
   (76.2 × 101.6 × 12.7 mm × 3.1 m)

Unequal leg

Angle or L

2-4 *Angle shapes and how they are stated.*

for details. Pipe may also be listed as follows: Standard is schedule 40, extra-strong is schedule 80, and double extra-strong is schedule 120.

If a designation is given on a structural drawing or blueprint such as 4" #40 blk. P.E., you would identify it as 4-inch, standard black pipe, with plain ends (no threads, but a 30-degree bevel on each end). If a designation is given as Gal 4" std. T and C, the proper identification is 4-inch(101.6-mm), galvanized, standard pipe, threaded, the threads protected by a cap (usually plastic).

Galvanized pipe is not generally preferred for fabricated structures. When welded or heated, the galvanized surfaces break down and give off zinc oxide, which is a very serious health hazard (see Chapter 9).

For both pipe and tubing, flat plate is cut to exact width. The name for this plate is *skelp*. The skelp is then rolled past forming dies

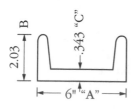

"A" Size of channel
"B" Size of flange
"C" Thickness of web
"L" Length of channel

Example
6" × 2.03 × 3.43 × 20'0"
152 × 5.1594 × 8.713 mm × 6.2 meters

2-5 *Standard channel shapes.*

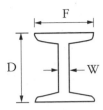
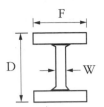

Taper flange
Standard I beam
Now M and S beams

Parallel flanges
Wide flange beams
Also **W-** listing

"D" Size-depth of shape
"F" Flange width
"W" Web thickness

6" × 6" × 20 lbs. × 20'0"
(152 × 152 mm × 9.0 kg × 6.2 m)

2-6 *I beam (now called M (mill) or S (standard) beam) and wide-flange beam.*

2-7 *Warehousing of wide-flange beam and pipe.*

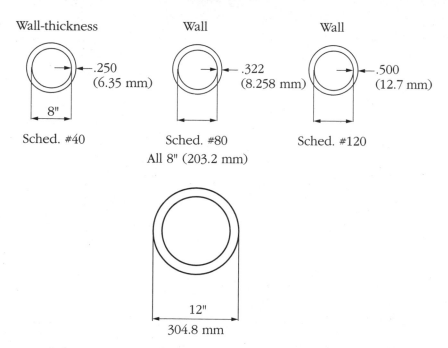

O.D. Measurement for all pipe over 12" (304.8 mm)

2-8 *The wall thicknesses of pipe and the changeover point from I.D. to O.D.*

until the edges meet. A surge of electricity is applied to the seam as pressure is held on the dies. The result is a welded hot-rolled pipe or tube. Figure 2-9 shows the details of square and rectangular tubing. The seam weld (weld symbols are explained in Chapter 8) is a form of resistance welding. While structural tees are not widely used in outside construction, they do show some resistance to shear and load factors. See Fig. 2-10 for these details.

Junior channel is now used more than *ship channel* because of its strength in relation to its weight. Some load tables list it as *light-weight channel.* See Fig. 2-11 for details and identification purposes. Ship and *car channels* are the same structurally and in dimension. Only the names change because of use. A draftsperson or engineer may use these terms. These last two channels are detailed in Fig. 2-12. As shown, standard and junior channel have tapered flanges, while ship channel has parallel flanges.

The less common shapes are generally for special use, and I mention them for identification purposes only. Z forms and modified Zs are used mainly for railroad boxcar manufacture. The ball and socket, thumb and finger, straight web, arch web, and 2-web piling forms are used mainly for dams and coffer dams.

Shown as
D   Depth of section
W   Width of section
T   Wall thickness
L   Length of tube

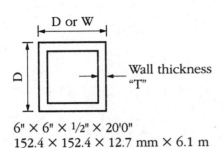

D or W

Wall thickness
"T"

6" × 6" × ¹/2" × 20'0"
152.4 × 152.4 × 12.7 mm × 6.1 m

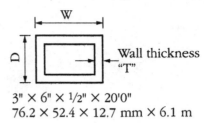

W

Wall thickness
"T"

3" × 6" × ¹/2" × 20'0"
76.2 × 52.4 × 12.7 mm × 6.1 m

2-9 *Structural shapes of tubing.*

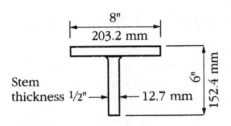

8"
203.2 mm

Stem
thickness ¹/2" → ← 12.7 mm

6"
152.4 mm

"T" section or structural "T"

Expressed as
S.T. 8" × 6" × ¹/2" × 44 lbs. per ft.
S.T. (203.2 × 152.4 × 12.7 mm)
× 20 kg for each 304.8 mm of
length.

2-10 *T shapes cut from wide-flange beam.*

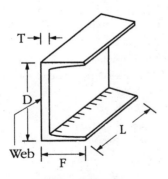

T →| |←

D

Web |←   →| ⟩⟨
        F

Junior channel
Identified by-JR-C
or JR]
Tapered flanges

D × F × T × L or D-X = WT/FT

D = Depth of] in inches
F = Width of flange in inches
T = Thickness of web in inches
L = Length in feet and/or inches

Expressed as
12" × 1¹/2" × ³/16" channel × 16'0"
or 12" JR] 12" × 1¹/2" × ³/16" × 16'0"
or 12" × 1¹/2" × ³/16" JR] × 16'0"
or 12" × 10.5 × 16'0"
or 304.8 × 38.1 × 4.7625 mm - JR]
× 4.9 m long

2-11 *Shape and size of junior channel.*

PARALLEL FLANGES

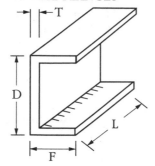

Ship-or-car
channel
identified as
ship or CAR]

D × F × T × L or D-X-wt./ft.

D = Depth of] in inches
F = Width of flange in inches
T = Thickness of web in inches
L = Length in feet and/or inches

Expressed as
8" × 3" × 7/16" Ship channel × 14'0"
or 8" ship] 8" × 3" × 7/16" × 14'0"
or 8" × 3" × 7/16" SHIP] × 14'0"
or 8" × 20. × 14'0" (20 = lbs. per lin. ft.)
or 203.2 × 76.2 × 11.1125 mm Ship]
4.25 m long

2-12 *Ship channel is a heavier section than either junior or standard channel.*

Since most structural shapes are not overstressed in compression, let's test them with other forms of stress. Shear is usually the villain in the failure of welded structures. Of course, it is often accompanied by vibratory stress. Any changing load factor can be deemed *vibratory stress* or *live load*. Snow melting on a structure or loaded trucks crossing a bridge both constitute live load.

These two stresses lead to a condition known as metal fatigue. One theory is that the inline grain structure of the rolling process is reversed and the crystals line up so the loading can cause the fracture of thin sections of individual crystals. If a truck frame cracks, the crack will be at a spring hanger or motor mount. Both of these areas are subject to both stresses mentioned. In observing any crack in a frame, you'll see proof of one item. The crack is never a straight line—it follows the path of least resistance. A small end of a crystal snaps because it lacks support. The dead air space between the grains lie directly below the crack, and if a path then opens to one side or the other, the crack then follows in that direction. Figure 2-13 shows possible crack lines.

Torsional stress is not normally a factor to be considered by fabricators. A car axle is good example of this type of sudden twisting load. The one stress we cannot protect against is impact loading. We cannot predict the load weight or the speed at impact.

We have delineated many structural shapes. Of course, many other shapes are available, but they are primarily used for other fields. Shafting and cold-finished stock is generally used in machine work. Reinforcing bar or rod is for strengthening concrete. Cast shapes may

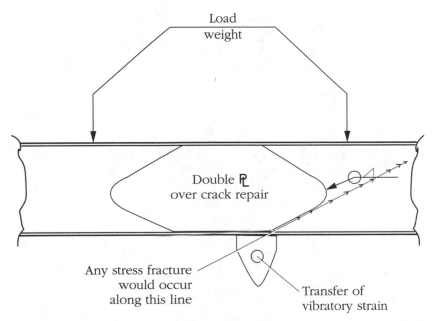

Load
weight

Double ℔
over crack repair

Any stress fracture
would occur
along this line

Transfer of
vibratory strain

2-13 *The fish plating of a fracture-prone fabricated item reduces the chance of failure.*

be used in all these fields. Figures 2-3, 2-14, and 2-15 show rebar formed to strengthen concrete bridge abutments and approaches.

The following tables and specifications supply any possible need for fabrication purposes. These shapes are described in Tables 2-2 through 2-9. Of course, all plates and shapes are made as close to

2-14 *Reinforcing rod, or rebar, in various shapes.* Farwest Steel

arwest offers full service rebar fabrication facilities. Using aSa® software, our estimators make accurate material take offs to insure precise bidding. Detailers use the aSa® CAD/Detailing System for speed, accuracy, and standardization of field drawings. Your CAD based structural drawings can be easily imported via modem or diskette to further improve construction performance.

FARWEST STEEL
Corporation

2-15 *Some exotic shapes of rebar for strengthening of concrete.* Farwest Steel

standard as possible. Sections are often sold as substandard and noted as such. This is true of pipe and tube to a greater degree because of methods of manufacturing. You will notice that extreme problems would arise if you attempted to butt together shapes as far out as the tolerances allow.

Table 2-7 shows data on camber. It may be a very desirable quality for fabricated structures, and beams having a depth of 24 inches and more can be ordered with specific camber. Since arch or camber, if turned up, would represent compression load until all camber was gone, it becomes apparent that failure could not occur until this point is reached and exceeded.

Table 2-2. Width and length tolerances for sheared plates (up to 1½ inches thick) and length tolerances for universal mill edge plates (up to 2½ inches thick).

Tolerances over Specified Width and Length for Thicknesses, inches, and Equivalent Weights Given

Widths	Lengths	To 3/8, excl. (To 15.3 lb. per sq. ft., excl.) Width	Length	3/8 to 5/8, excl. (15.3 to 25.5 lb. per sq. ft., excl.) Width	Length	5/8 to 1, excl. (25.5 to 40.8 lb. per sq. ft., excl.) Width	Length	1 to 2, incl.* (40.8 to 81.6 lb. per sq. ft., incl.) Width	Length
To 60, excl.	To 120, excl.	3/8	1/2	7/16	5/8	1/2	3/4	5/8	1
60 to 84, excl.		7/16	5/8	1/2	11/16	5/8	7/8	3/4	1
84 to 108, excl.		1/2	3/4	5/8	7/8	3/4	1	1	1-1/8
108 and over		5/8	7/8	3/4	1	7/8	1-1/8	1-1/8	1-1/4
To 60, excl.	120 to 240, excl.	3/8	3/4	1/2	7/8	5/8	1	3/4	1-1/8
60 to 84, excl.		1/2	3/4	5/8	7/8	3/4	1	7/8	1-1/4
84 to 108, excl.		9/16	7/8	11/16	15/16	13/16	1-1/8	1	1-3/8
108 and over		5/8	1	3/4	1-1/8	7/8	1-1/4	1-1/8	1-3/8
To 60, excl.	240 to 360, excl.	3/8	1	1/2	1-1/8	5/8	1-1/4	3/4	1-1/2
60 to 84, excl.		1/2	1	5/8	1-1/8	3/4	1-1/4	7/8	1-1/2
84 to 108, excl.		9/16	1	11/16	1-1/8	7/8	1-3/8	1	1-1/2
108 and over		11/16	1-1/8	7/8	1-1/4	1	1-3/8	1-1/4	1-3/4
To 60, excl.	360 to 480, excl.	7/16	1-1/8	1/2	1-1/4	5/8	1-3/8	3/4	1-5/8
60 to 84, excl.		1/2	1-1/4	5/8	1-3/8	3/4	1-1/2	7/8	1-5/8
84 to 108, excl.		9/16	1-1/4	3/4	1-3/8	7/8	1-1/2	1	1-7/8
108 and over		3/4	1-3/8	7/8	1-1/2	1	1-5/8	1-1/4	1-7/8
To 60, excl.	480 to 600, excl.	7/16	1-1/4	1/2	1-1/2	5/8	1-5/8	3/4	1-7/8
60 to 84, excl.		1/2	1-3/8	5/8	1-1/2	3/4	1-5/8	7/8	1-7/8
84 to 108, excl.		5/8	1-3/8	3/4	1-1/2	7/8	1-5/8	1	1-7/8
108 and over		3/4	1-1/2	7/8	1-5/8	1	1-3/4	1-1/4	1-7/8
To 60, excl.	600 to 720, excl.	1/2	1-3/4	5/8	1-7/8	3/4	1-7/8	7/8	2-1/4
60 to 84, excl.		5/8	1-3/4	3/4	1-7/8	7/8	1-7/8	7/8	2-1/4
84 to 108, excl.		5/8	1-3/4	3/4	1-7/8	7/8	1-7/8	1-1/8	2-1/4
108 and over		7/8	1-3/4	1	2	1-1/8	2-1/4	1-1/4	2-1/2
To 60, excl.	720 and over	9/16	2	3/4	2-1/8	7/8	2-1/4	1	2-3/4
60 to 84, excl.		3/4	2	7/8	2-1/8	1	2-1/4	1-1/8	2-3/4
84 to 108, excl.		3/4	2	7/8	2-1/8	1	2-1/4	1-1/4	2-3/4
108 and over		1	2	1-1/8	2-3/8	1-1/4	2-1/2	1-3/8	3

Tolerance under specified width and length, 1/4 in.

* Length tolerances apply also to U.M. plates up to 12 inches in width for thicknesses over 2 to 2½ in. incl.

American Iron and Steel Institute

Table 2-3. Tolerances for overweight and underweight structural sections

Tolerance for the calculated or specified weight
is plus or minus 2.5 per cent.

TOLERANCES FOR STANDARD BEAMS, STANDARD MILL H-BEAMS, CHANNELS

Section	Specified Size, in.	A Depthª		B Flange Width		T + T' Out-of-Square Per Inch of B, in.
		Over, in.	Under, in.	Over, in.	Under, in.	
Standard Beams	3 to 7, incl.	3/32	1/16	1/8	1/8	1/32
	Over 7 to 14, incl.	1/8	3/32	5/32	5/32	1/32
	Over 14 to 24, incl.	3/16	1/8	3/16	3/16	1/32
Standard Mill H-Beams	4	3/32	1/16	1/8	1/8	1/32
	5	3/32	1/16	5/32	5/32	1/32
	6 and 8	1/8	3/32	3/16	3/16	1/32
Channels	3 to 7, incl.	3/32	1/16	1/8	1/8	1/32
	Over 7 to 14, incl.	1/8	3/32	1/8	5/32	1/32
	Over 14	3/16	1/8	1/8	3/16	1/32

ª A is measured at center line of web for beams; and at back of web for channels.

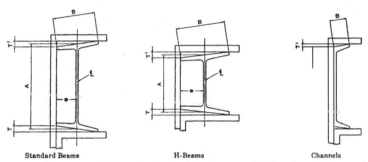

Standard Beams	H-Beams	Channels
⁕Back of Square and center line of Web to be Parallel when Measuring "Out-of-Square."		T + T' applies when flanges of channels are toed in or out regardless of the direction of toeing.

Standard Beams, Standard Mill H-Beams, Channels

American Iron and Steel Institute

Availability, aesthetics, and pricing

Any steel catalog will give you a rough idea of what a rolling mill can
supply. Due to transportation problems, plates are not often supplied
as large as those listed in Table 1-4. Since transport is usually by rail,
size is not a factor in shapes. Standard W beams are 36 inches high
and 300 pounds per feet by 60 feet long, but 52 inches by 620 pounds
by 100 feet long can be supplied. At any point above 24 inches by
100 pounds in a desired length, it may be more economical to fabricate the beams. Do not attempt to fabricate channel or angle.
Warpage and stresses will cause severe problems in every case. Angles and channels, however, can be used to provide box sections, as
shown in Fig. 2-16.

The fabricated sections have obvious advantages. The box channel section can be fabricated in larger sizes than rectangular tubes
produced by the rolling mills. This section can also be cambered

Table 2-4. Tolerances for angles, bulb angles, rolled tees, and zees

Section	Specified Size, in.	A Depth Over, in.	A Depth Under, in.	B Flange Width or Length of Leg Over, in.	B Flange Width or Length of Leg Under, in.	T Out-of-Square Per Inch of B, in.	E Web Off Center, in.
Angles[a]	3 to 4, incl.			1/8	3/32	3/128[b]	
	Over 4 to 6, incl.			1/8	1/8	3/128[b]	
	Over 6			3/16	1/8	3/128[b]	
Bulb Angles	(Depth) 3 to 4, incl.	1/8	1/16	1/8	3/32	3/128[b]	
	Over 4 to 6, incl.	1/8	1/16	1/8	1/8	3/128[b]	
	Over 6	1/8	1/16	1/8	1/8	3/128[b]	
Rolled Tees	Stem or Flange 5 and under	3/32	1/16	1/8	1/8	1/32	3/32 Max.
	Stem or Flange over 5 to 7, incl.	3/32	1/16	1/8	1/8	1/32	1/8 Max.
Zees	3 to 4, incl.	1/8	1/16	1/8	3/32	3/128[b]	
	Over 4 to 6, incl.	1/8	1/16	1/8	1/8	3/128[b]	

[a]For unequal leg angles, longer leg determines classification.
[b]3/128 inch per inch = 1-1/2°.

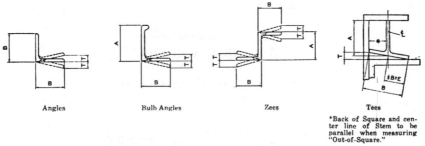

Angles　　　Bulb Angles　　　Zees　　　Tees

*Back of Square and center line of Stem to be parallel when measuring "Out-of-Square."

Fig. 7.　Angles, Bulb Angles, Zees, Rolled Tees

American Iron and Steel Institute

during the welding process if desired. The box angle section can be fabricated in sizes and wall thickness not supplied by the rolling mills, but care must be exercised in the welding process so that sweep is not an excessive factor.

The W beam and flat plate boxed section adds transverse and shear strength to the section and additional resistance to side thrust. The W beam and flat plate section with plate added to the top and bottom flanges greatly reduces the failure of section in shear and tensile load situations. Cambering can be a part of the welding process. These combination sections are shown in Fig. 2-17.

Table 2-5. Length tolerances for all standard sections

All Standard Sections	Tolerances, in., for Specified Length for Lengths Given								
	To 30 ft. Incl.		Over 30 ft. to 40 ft., Incl.		Over 40 ft. to 50 ft., Incl.		Over 50 ft. to 65 ft., Incl.		Over 65 ft.
	Over	Under	Over	Under	Over	Under	Over	Under	Over	Under
	1/2	1/4	3/4	1/4	1	1/4	1-1/8	1/4	1-1/4	1/4

American Iron and Steel Institute

Table 2-6.
Tolerances for end out-of-square
structural sections

Beams, Channels . . . . . .⎫ $\frac{1}{64}$ in. per inch of depth.
Standard Mill H-Beams . .⎭

Angles[a] . . . . $\frac{3}{128}$ in. per inch of leg length or $1\frac{1}{2}°$.

Bulb Angles . . . . $\frac{3}{128}$ in. per inch of depth or $1\frac{1}{2}°$.

Rolled Tees[a] . . . . $\frac{1}{64}$ in. per inch of flange or stem.

Zees . . . . $\frac{3}{128}$ in. per inch of sum of both flange lengths.

[a]Tolerances for ends out-of-square are determined on the longer members of the section.

Camber tolerance
all standard sections

(Camber as shown in Fig. 8 is measured over the entire length of the section.)

$$\frac{1}{8}\text{ in.} \times \frac{\text{Number of feet of total length}}{5}$$

Sweep. Due to the extreme variations in sweep of standard beams and channels, allowances for sweep should be established in each instance.
American Iron and Steel Institute

The double-angle strut section is often employed in the fabrication of trusses for large structures (Fig. 2-18). Besides doubling the yield point strength of a single angle, the resistance to side thrust is more than doubled, and when the flat plate is extended past the legs of the angle, other shapes may be welded to it without setting up metallurgical change in the angles themselves.

Availability is usually limited by transportation methods and a steady demand factor. Suppliers cannot and will not keep stock that does not have a reasonable turnover rate. Inventory taxes, storage space, and ease of handling affect the supply of steel in any local warehouse. If rail transport is available, any rolling mill will supply any section within the limits stated if it is ordered in carload lots, that is, 138,000 pounds or more of any one section. While this weight seems to be more than normally used, it becomes more practical if considered as 5 plates that are 3 inches thick by 10 feet wide by 20 feet

Table 2-7. Tolerances for wide-flange sections

Section Nominal Size, in.	A Depth		B Flange Width		T + T' Flanges Out-Of-Square, in.	E Web Off Center, in.	C Max. Depth at any Cross Sect.
	Over Theoretical, in.	Under Theoretical, in.	Over Theoretical, in.	Under Theoretical, in.			Over Theoretical Depth, in.
Up to 12, incl.	1/8	1/8	1/4	3/16	1/4 Max.	3/16 Max.	1/4
Over 12	1/8	1/8	1/4	3/16	5/16 Max.	3/16 Max.	1/4

A is measured at center line of web.
B is measured parallel to flange.
C is measured parallel to web.

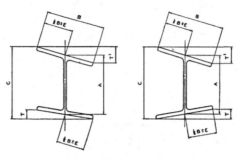

Wide Flange Sections

American Iron and Steel Institute

long and one plate that is 16 feet long or a 36-inch W beam at 300
pounds per foot (460 feet total length). A manufacturer's directory and
a few phone calls can often fill the needs of the fabricator.

Aesthetics

Structural steel has a stark, utilitarian aspect, and with good reason—
a skeleton doesn't look good either. The square corners and jagged
edges serve a purpose, and no other material does as well.

The use of combination shapes does somewhat soften the outline
of a structure. Round, square, and rectangular shapes are often the
first choice of the designer. They combine excellent strength without

Table 2-8. Length tolerances and ends out-of-square tolerances for wide-flange sections

Nominal Depth, in.	Tolerances, in., for Specified Length for Lengths Given			
	To 30 ft., incl.		Over 30 ft.	
	Over	Under	Over	Under
Beams up to 24, incl.	3/8	3/8	3/8 plus 1/16 for each additional 5 ft. or fraction thereof	3/8
Beams over 24 and all Columns	1/2	1/2	1/2 plus 1/16 for each additional 5 ft. or fraction thereof	1/2

When Wide Flange Sections are used as Bearing Piles, the length tolerance is plus 5 in. and minus 0 in.

Ends Out-of Square

1/64 in. per inch of depth, or of flange width if it is greater than depth.

American Iron and Steel Institute

Table 2-9. Tolerances for camber and sweep
of wide flange sections

$$\frac{1}{8} \text{ in.} \times \frac{\text{number of feet of total length}^a}{10}$$

[a]Sections with flange width less than 6 in., tolerance for

$$\text{sweep} = \frac{1}{8} \times \frac{\text{number of feet of total length}}{5}$$

(Camber and sweep, shown in Fig. 10, are measured over the entire length of the section.)

When certain sections[b] with flange width approximately equal to depth are specified on order as columns:

Lengths to 45 ft., inclusive:

$$\frac{1}{8} \text{ in.} \times \frac{\text{number of feet of total length}}{10} \text{ but not over } \frac{3}{8} \text{ in.}$$

Lengths over 45 ft.:

$$\frac{3}{8} \text{ in.} + \frac{1}{8} \text{ in.} \times \frac{\text{number of feet of total length, minus 45}}{10}$$

[b]Applies only to:

> 8 in. deep sections 31 lb./ft. and heavier
>
> 10 in. deep sections 49 lb./ft. and heavier
>
> 12 in. deep sections 65 lb./ft. and heavier
>
> 14 in. deep sections 78 lb./ft. and heavier

This manual section does not contain tolerance for other sections which are specified as columns.

American Iron and Steel Institute

leaving raw edges. If there is a need for dead air space, contained or concealed equipment, or the movement of liquids throughout a structure, then these are the architect's first choice.

Facia is available not only as a cover for structural steel but may indeed be a part of that structure. Cast and fabricated metal shapes are used for strength, durability, and beauty on many large buildings. Any good ornamental iron and steel supply house will have catalogs of materials for these purposes. Privacy screens are available for

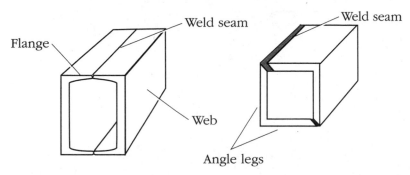

Channel box section fabricated   Angle box section fabricated

2-16 *The structural strengthening of composite shapes.*

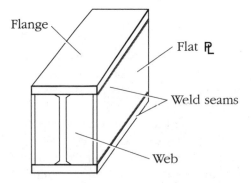

Combination **W-** beam and flat plate

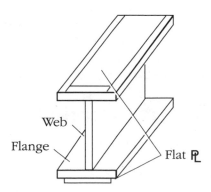

Combination **W-** beam with plate added

2-17 *Beams are made stronger by the addition of flat bar.*

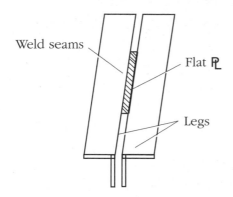

Weld seams

Flat ℞

Legs

Combination section angle
and flat plate fabricated
identified as double angle struts

2-18 *Double-strut construction with the addition of flat tie plates.*

balconies, louvers for windows, as are spiraled or curved stairways, trellised or suspended walkways, arches, and bridges.

Pricing

Mild or low carbon steel has increased in price since 1940 in a rather natural way. As labor and transportation costs moved upward, so did the making and distributing of steel to places of use. Steel that sold for roughly 8 cents a pound in 1965 made a sharp jump to around 35 cents by 1975 and stayed at about that point since that time. Check prices with a local warehouse before ordering.

No new steel mills have been built to produce "new steel" in the United States for more than 20 years. A modern blast furnace is not cost effective in view of the price of imported steel after it is subsidized by countries to enhance their own import positions.

Small companies have done well using reclaimed steel and electric arc processes; however, many structural shapes are beyond their capabilities. The pricing of low-alloy, high-tensile steels and alloys, high carbon, and stainless steels are another matter. The least expensive of these will cost twice as much as mild steel. Those costs are detailed later.

Project 3: Structural shapes

The cutting and welding of structural steel shape is a first requirement for all fabricators/welders. From the points shown in Fig. 2-19, cut and fit one channel-to-channel joint using good practice techniques. Then cut and fit one wide flange beam to another wide beam of similar dimensions. Make sure that the shapes are at right angles to each other in both cases. Check for 90-degree correctness before and after

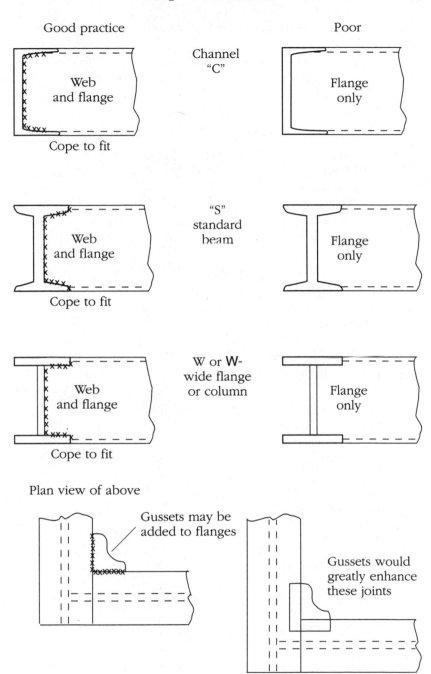

Good practice   **Poor**

Web and flange | Channel "C" | Flange only
Cope to fit

Web and flange | "S" standard beam | Flange only
Cope to fit

Web and flange | W or **W**- wide flange or column | Flange only
Cope to fit

Plan view of above

Gussets may be added to flanges

Gussets would greatly enhance these joints

2-19 *Correct view of fitting shapes.*

welding, using either a combination square or framing square. If you wish to add gussets, and it would not complicate the welding of the joint, then tack weld both the shapes and the gussets. The gussets will help hold the joint to true 90 degrees during the welding and cooling process. Be your own best critic.

Quick check

- Was your layout correct, and did you check it twice?
- Were your cuts almost slag free?
- Did you check again for angle?
- Was your welding process the best available?
- Are your weld beads the correct size and shape?
- Are you satisfied with the completed joints?

Project 4: Pipe joints

Using 4-inch schedule 40, black, plain end pipe, cut and weld a 90-degree joint as shown in Fig. 2-20. The metal-cutting band saw can be set for 45 degrees and will accommodate 4-inch pipes.

You may want to weld the joint. To achieve maximum joint strength, you must bevel the pipe using a power grinder or torch. Space the pieces $\frac{1}{16}$ inch or $\frac{3}{32}$ inch apart for a root gap opening that ensures complete melt-through, which means that you have a small uniform bead that has penetrated to the inside of the pipe.

Questions for study

1. What is the difference between standard and ship channel?
2. What is the difference between standard and junior channel?
3. What is the largest size of angle presently available in most areas of the United States?
4. Does weight or shape have the most effect on structural strength?
5. Does channel or mill beam resist side stress to a greater degree?
6. Is weight an identifying factor in nominal sizes of angle?
7. What letters identify standard and mill beam?
8. What is the major difference between standard and wide-flange beams?
9. What is the largest standard beam you could obtain in your area? How would you order it?
10. Is a "trial run" a common mill practice in the making of steel?

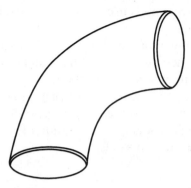

Long-radius elbow with 30° beveled ends for welds
90° elbow for pressure piping

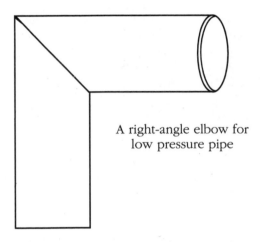

A right-angle elbow for
low pressure pipe

90° fabricated pipe elbow for structural use
or for fluids at very low pressure

2-20 *A right-angle elbow for low-pressure pipe.*

11. In mill practice, are shape and section the same?
12. Is preheat necessary in the welding of mild steel? Is thickness a factor to be considered in the welding process?
13. If the extremes of mill tolerance on two wide flange beams were encountered, what problems would occur in the butt welding of one beam to the other?
14. If W is the common identity symbol for wide-flange beam, what other symbol may replace it?

15. Resistance welding is used in the manufacture of which structural shapes?
16. Is repair allowed on rolled structural shapes?
17. If there is a need to measure the thickness of a rolled or sheared plate, where should the measurement be taken?
18. To what extent is squareness of plate guaranteed?
19. About how much has the price of steel increased in 20 years?
20. Mild steel will cost at least how much less than low-alloy, high-tensile-strength steel?

3

Low-alloy, high-tensile steels

Low-alloy, high-tensile-strength steels, grades, metallurgy, weights, and abrasion factors are all a part of this chapter. These steels have achieved great success with the fabrication industry because of their superior weight reduction and abrasion factors.

The United States Steel Corporation has a very fine line of these steels. Table 3-1 shows the elements (ingredients) for one of these steels. Note the small amounts of any element to the total percentage of the steel. Pay particular attention to the tensile strength of the steel. The material under 2½ inches in thickness quenched more quickly and the grain (crystal structure) reacted to give an increased tensile-strength advantage.

Low-alloy, high-strength steel is at least twice the strength of A-36 mild steel. You see that this steel can achieve a better weight-to-strength factor. If you produce a trailer or other fabricated item that must cross a state or federal weight scale device, half the weight of the item can now be changed to payload. It also has a 6-to-1 abrasive resistance factor, a distinct advantage over aluminum when any structural wear is involved.

The fabrication data of these steels always lists cold bend properties. This simply means *don't* hot bend this material. Any heat above 1100 degrees F makes the loss of high strength certain. The steel then has the strength of low carbon steel. Many other companies produce similar steels that are excellent, and the fabrication data remains the same.

Other than total cold-bend forming, the other problems with these steels are that you must use low-hydrogen filler metals, paying attention to storage and welding conditions and, in some cases, stress relieving. Tables 3-2 and 3-3 provide valid information on electrode use. If you need to run tests in which guided bends are required, you may need to use 70xx or 80xx class electrodes, particularly on root passes.

Table 3-1. Chemical composition* (for information only)

C	Mn	P† Max.	S† Max.	Si	Ni	Cr	Mo	V	Cu	B
.10/.20	.60/1.00	.040	.050	.15/.35	.70/1.00	.40/.80	.40/.60	.03/.10	.15/.50	.002/006

*U.S. Patent No. 2,586,042

†For qualities higher than Regular Quality the phosphorous and sulphur limits are lowered to conform to ASTM Standards.

Additional Information For Engineering Guidance:

Heat Treatment

USS "T-1" steel is water quenched from 1650/1750°F and tempered at 1100/1275°F.

Modulus of Elasticity

In tension . . . . .approx. 30,000,000 psi
In compression . . . . .approx. 30,000,000 psi

Coefficient of Expansion

7.74×10^{-6} inches per inch per °F in the range of 70° to 1300°F.

Weldability

Joint efficiency—with AWS 11015, 12015 or
 equivalent electrodes 100%
Joint efficiency—automatic welding 100%
Kinzel transition temperature—
 welded ½ and 1" plate specimens...minus 40/90°F
Maximum hardness—heat affected zone . . . 410 DPH
Minimum hardness—heat affected zone . . . 260 DPH

Fatigue Strength

Rotating beam endurance limit—
 polished specimen . . . . .67,000 psi
Pulsating fatigue endurance limit—
 unwelded (surface as-rolled) . . . . 50,000 psi

Atmospheric Corrosion Resistance

Four times that of structural carbon steel.

High Temperature Strength

Creep rupture strength at 900°F—three times that of carbon steel and equal to conventional ½% Cr—½% Mo steel.

Cold Bend Properties

Cold Bend	.1875" to .249", Incl.	¼" to 1", Incl.	Over 1" to 2", Incl.	Over 2" to 4", Incl.
Transverse Test	90°D=4T	180°D=2T	180°D=3T	180°D=4T
Longitudinal Test	180°D=2T	180°D=2T	180°D=3T	180°D=4T

*Longitudinal bend tests are made except when Flange or Firebox Quality is specified in which case transverse bend tests are made.

You should understand that high yield-point weld deposits are beyond the fracture point of many filler metals. This in no way reflects on the competency of the welding operator. The boiler and pressure vessel codes recognize this fact and qualify the *operator's proficiency.*

Figure 3-1 clearly shows the difference in test radius mandrels, but it does not show that 70xx and 80xx data might be included to your advantage. This is not contrary to test code. You will revert back to 90xx, 110xx, and 120x for the actual welding of the fabricated items.

Many gas metal arc welding (GMAW) filler metals are available and are improving constantly. If you are welding on the low-alloy, high-strength steels on thickness above $\frac{5}{16}$-inch (7.937 mm), you may choose large-diameter wires—$\frac{3}{32}$-inch (2.381-mm) or even $\frac{7}{64}$-inch (2.778-mm) cored wire. The heat input to steels will not exceed safe practice because of speed of travel, which means increased production without loss of quality. If work can be positioned for true flat or horizontal filled welding, the travel speed and deposition rate of filler metal will increase in direct proportion to the amount of amperage and voltage used. If welding with a $\frac{3}{16}$-inch (4.7625-mm) electrode at 180 amps gives you 18 to 24 inches per minute and $\frac{7}{64}$-inch cored wire is deposited at 475 amps, the same size bead can be laid down at about 60 inches per minute.

The most important fact about any of these steels is that they are tempered, and their metallurgical makeup reacts to heating. Keep this fact in mind during all cutting, welding, and forming operations. Stress relieving up to 1100 degrees F (about 590 degrees C) saves many welds in which residual stress was found to be the culprit. The rolling (direction of rolling), shearing, forming, and torch cutting were of course done before the welding process ever entered the picture. It has been found that rolling the steels in the lengthening process requires further rolling in a transverse direction to change the grain pattern (crystal structure). If you order steel, make sure that if it needs forming in a press brake that you use radius-forming dies. The minimum radius should be twice the thickness up to 1 inch thick and three times the thickness up to 2 inches thick. Rather than the sharp 90-degree (right angle bend) dies, also understand that even with radiused bends, cracking may occur at the sharpest point of directional change. Therefore, simply order double-rolled material if in doubt.

The American Society of Mechanical Engineers (ASME) now states that all such steel used in boilers and pressure vessels must be stress-relieved. T-1 steel and its industry-wide counterparts receive a great deal of attention in this text because they need special fabrication and welding techniques to realize their full potential.

Table 3-2. Results of tests of longitudinally butt-welded tension specimens of "T-1" steel

Electrode Type	Plate Thickness	Heat No.	Condition of Weld	Yield Strength (.02% Offset), psi*	Tensile Strength, psi*	Elong. in 8 in., %*	Elong. in 2 in., %*
E12015	½"	36S462	As-welded	102,900	120,200	11.2	.....
E12015	1"	29S144	As-welded	101,400	119,900	11.3	27.7
E9015	1"	32U029	As-welded	102,200	112,400	14.7	32.3
E9015	1"	32U029	Stress relieved (1100°F)	102,100	111,100	16.0	33.3
None (base metal)	½"	74L236	....	115,000†	123,000†	11.0†	25.0†

*Average of three values.
†Typical value.

Table 3-3. Results of tests transversely butt-welded tension specimens of "T-1" steel

Electrode Type	Plate Thickness	Heat No.	Condition of Weld	Reinforce- ments	Tensile Strength, psi*	Location of failure	Joint Efficency, %†
E12015	½"	36S462	As-welded	Off	118,700	Base metal	100
E12015	1"	29S144	As-welded	Off	115,400	Base metal	100
E9015	1"	73U115	As-welded	Off	117,100	Weld metal	93.8
E9015	1"	73U115	Stress relieved (1100°F)	Off	113,900	Weld metal	91.4
E9015	1"	32U029	As-welded	On	121,500	Base metal	100
E9015	1"	32U029	Stress relieved (1100°F)	On	117,250	Weld metal	94
E12015	½"	74L236	As-welded	Off	125,400	Base metal	100
E9015	½"	74L236	Stress relieved (1125°F)	Off	119,000	Weld metal	95

*Average of three values.   With express permission of U.S.S. Steel
† Based on actual strength of unwelded plate.

Table 3-4 gives welding filler metal chemical composition and stress-relieving and interpass temperatures. For most boiler work the material need not be held more than 12 hours for stress relieving and should not exceed 1100 degrees F (approximately 580 degrees C). It should be blanketed with insulation and allowed to cool slowly down to 200 degrees F (approximately 93 degrees C). The interpass temperature is noted as 100 degrees F (about 38 degrees C) and should not exceed 250 degrees F (120 degrees C).

The use of manganese, nickel, chromium, and molybdenum together means that it hardens well on quenching at low temperature without becoming a clinched grain structure (martensite, found in high carbon steels).

Figure 3-2 shows penstock (pipeline) fabricated from T-1 rolled to a perfectly round shape. You can see that it is used to convey water from an upper level down to turbine generators at a power house. The water attains column pressure, and, if necked down, the nozzle force will turn propeller-like blades that generate electricity. The savings from such use could come from the following factors: the same or better strength using half the weight, the cost of welding operations almost cut in two, no cost for preheat or stress-relieving, reduced costs for foundation materials, fittings and handling, and shipping and erecting. While it may not save half the cost, it should save more than one-third and up to almost half the total cost! Do not conclude from this that all pipelines or tanks should be built from such materials.

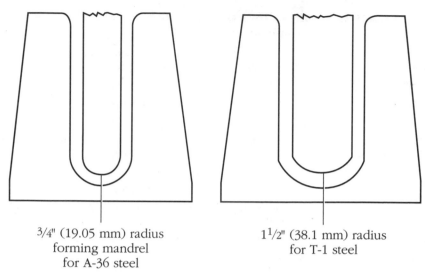

3/4" (19.05 mm) radius
forming mandrel
for A-36 steel

1¹/₂" (38.1 mm) radius
for T-1 steel

3-1. *Exact dimensions for making test dies for low-alloy, high-strength steels.*

Storage tanks that will contain 100,000 to 500,000 gallons (386,000 to 1,930,000 liters) of liquid (water or petroleum products) will normally use mild steel plate only ⁵⁄₁₆-inch to ⅜-inch (7.937-mm to 9.525-mm) thick, and that perhaps only for the first tier of plates, graduating down in thickness as additional height is added. This means that few, if any, of the cost factors mentioned, would even be considered for fabrication. Each fabricated item should be evaluated on such things.

Abrasive-resisting (AR) plates were first thought to mean armor plate by fabricators after WWII, and perhaps some plates sold at that time were left over from the war stockpiles. The following data gives you some metallurgical facts and uses for such steels. An analysis of chemical content is as follows: carbon 0.35 to 0.5 percent, manganese 1.50 to 2 percent, phosphorous 0.05 percent maximum, sulfur 0.005 percent maximum, and silicon 0.15 to 0.3 percent. As you can see, this steel is medium to high carbon with a Brinell hardness of 200 to 250 (15 to 24 on the Rockwell C scale). It has a rolled tensile strength of 100,000 to 125,000 pounds per square inch. The abrasion factor is about six times that of mild steel. Use it for conveyor bottoms, chutes, tubes, ductwork, or places where wear is a prime consideration. If concentrations of sulfur and or phosphorus become a welding problem, you might consider butter passes (the laying down of a weld bead at the point or points of joining dissimilar steels to each other). Run a pass on first one plate and then a pass on the other plate. This makes a composite mixture of each steel and weld metal without the *strain* of dissimilar metal cracking. Now weld the two butter passes together. You are now welding new weld metal to new weld metal without the problems of different cooling rates, carbon content, or metal shrinkage. This weld procedure is often used on dissimilar stainless steels as well.

Bethlehem, Republic, and United States steel companies are leaders in the research and production of specialty steels in the low-alloy, high-strength steel field. Other companies have entered the market with their own brand names, and they often branch into cold-roll and tool-steel materials. The United States does not offer competitive prices with Japan, Germany, and emerging industrial nations on most mild steel structural shapes.

Cor-ten is another grade of steel in the low-alloy category. It is another U.S. Steel trade-name product. It has some excellent properties and is offered at a lower price than T-1 and its counterparts. It is a low carbon steel with enough nickel, chrome, and manganese to make it workable and weldable, and it is 4 to 6 times as corrosion-resistant as mild steel. Its tensile strength is about 1½ times that of mild steel. By workable, we mean that it drills, punches, and cold-forms almost the same as mild steel. While you could use the radius dies that you used for T-1 in the press brake, you might also get by with standard mild steel 90-degree bend dies. The welding of Cor-ten and other companies' trade-name products with similar elements percentages and properties still requires the use of low-hydrogen filler metals. You may want to use 70xx and 80xx fillers for increased ductility.

When we examine structural materials or fabricated members, certain facts become evident. Failure is always caused, and while it is divided into two distinct types, both may be present in certain types of fractures.

The first type of failure is *ductile* and is identified by necking along the crack line and the area adjacent to it. The grain structure will be elongated and may often lie in layers parallel to the crack. *Metal fatigue* may exhibit this type of fracture. A case in point would be the truck frame in Fig. 2-13. The cause would in part be attributed to vibratory stress. Fracture occurs at the motor mount area or where the load on the main members transfers to the axles (springhangers). The strengthening of the frame at these points would reduce the probability of failure and lengthen the life of the unit. Doubler or fish plates would be very effective. These plates, when welded to a frame either as a repair or insurance factor, add weight to the frame and reduce the payload. Further care must be taken in design and welding. Lines of force should be transverse to the length of the members, and sharp lines

Table 3-4. Chemical composition and mechanical properties of weld metals deposited by high strength electrodes

AWS Class	Comp. Type											
E9015	Mn-Mo	0.07	1.45	0.025	0.024	0.39	0.06	0.08	0.35	0.02	0.14	0.013
E11016	Mn-Ni-Mo	0.08	1.02	0.015	0.030	0.28	3.73	0.01	0.63	0.005	0.11	0.014
E12015	Ni-Mo-V	0.06	1.12	0.011	0.018	0.40	2.11	0.11	0.95	0.21	0.11	0.005

All-weld-metal tensile Properties (0.505" Diam. Specimen)

AWS Class	Comp. Type	Condition	Interpass Temp., °F	Yield Strength (0.2% Offset), psi	Tensile Strength, psi	Elong. in 2 in., %	Red. of area, %
E9015	Mn-Mo	As-welded	100	88,800	95,700	16.3	28.1
E9015	Mn-Mo	As-welded	Rising	61,750	84,600	29.0	62.5
E9015	Mn-Mo	Stress relieved (1100°F)	100	87,250	94,200	23.5	55.7
E9015	Mn-Mo	Stress relieved (1100°F)	Rising	67,500	82,500	26.8	63.1
E11016	Mn-Ni-Mo	As-welded	100	119,300	124,400	18.5	54.5
E11016	Mn-Ni-Mo	Stress relieved (1100°F)	100	106,200	113,400	21.2	58.6
E12015	Ni-Mo-V	As-welded	100	113,300	129,200	20.5	58.7
E12015	Ni-Mo-V	Stress relieved (1100°F)	100	116,000	128,700	18.8	59.4

Charpy-keyhole Impact Properties

AWS Class	Comp. Type	Condition	Ductility-transition Temp., °F
E9015	Mn-Mo	As-welded	minus 155
E9015	Mn-Mo	Stress relieved (1100°F)	minus 83
E11016	Mn-Ni-Mo	As-welded	minus 166
E11016	Mn-Ni-Mo	Stress relieved (1100°F)	minus 90
E12015	Ni-Mo-V	As-welded	minus 78
E12015	Ni-Mo-V	Stress relieved (1100°F)	130 above zero

3-2. *Exposed penstock used as a pipeline.* U.S. Steel

should be rounded. The same factors should be considered in lengthening or shortening any similarly loaded members. Welding should be
backstepped and all craters filled. Welds should never end at the edge
of a member and except for crack repair should never approximate a
line perpendicular to the frame.

The second type of fracture is *brittle fracture*. A combination of
factors may be said to contribute to such a metal failure:

- Rapid and/or extreme temperature drop.
- Sudden impact or torsional loading.
- A notch or notch effect, usually due to improper design
- Change in crystal structure of medium or high carbon and
  low-alloy steels (low-alloys exceeding the critical tempering
  range around 1300 degrees F in the fabrication or welding
  processes) due to carbide precipitation along grain
  boundaries and lack of heat treatment.
- Failure to understand the elongation factor in steel fabrication
  (each square inch of steel expands 0.00006 inch for each
  degree of heat applied).

A piece of steel 10 feet square heated 100 degrees F would expand 2.4 inches. If we restrain the steel on all four sides and weld a
level butt joint seam at its center, what happens as the steel cools and
attempts to shrink?

In actual practice, I have seen a deck seam on a ship split open
after being production-tacked for welding (4-inch tacks per foot) and

show a gap of 2¼ inches from simple contraction as it cooled. The sun's temperature probably exceeded 120 degrees F at 3 p.m., and the welds snapped at 10 p.m. The cooling was undoubtedly accelerated by the air as it moved from the water at the shipyard. The foregoing statement should prevent you from welding in direct precipitation. The melting point of mild steel is approximately 2850 degrees F, and the sudden cooling of any weld will set up a change in grain structure as well as extreme stress due to shrinkage. I do not recommend the quenching of any fabricated steel structure that has been welded. Any welded plate coupon, as shown in Fig. 1-8, will show fracture if the steel has an elongation factor below 18 percent in 2 inches or if the coupon is quenched. While the root pass and subsequent passes up to the top pass (cap) will show "normalized" grain structure, the cap will show a refined crystal pattern and embrittlements.

Questions for study

1. What does Table 3-1 tell you?
2. What is the importance of this information?
3. What makes thin plates higher in tensile strength than thick plates?
4. Are these steels stronger in equal-weight sections than mild steel?
5. If they are weaker than mild steel, how much weaker? If stronger, how much stronger?
6. Can you ever justify using a more expensive steel on a fabricated product?
7. If a welding test coupon fails, is it always the fault of the welder?
8. Would you generally use heat when bending these steels?
9. Are 6010 electrodes a good choice for the welding of these steels?
10. What are the advantages of using cored wires in welding these steels?
11. Would you sometimes stress-relieve this class of steels?
12. What kind of radius would you recommend for a 90-degree bend?
13. If you worked with this class of steels on a pressure vessel, what kind of interpass temperature should you not exceed?
14. What is a penstock?
15. What do you usually store in a round storage tank?
16. Is the correct name of AR plate armor plate?

17. In AR plate, is the carbon content higher or lower than that of T-1 steel?

18. What does the Brinell factor tell you about a steel?

19. Where might you prefer using AR plate over mild steel, and why?

20. Is the Cor-ten class of steels higher in tensile strength than the T-1 class of steels?

4

Shapes, availability, and prices

Check with your local warehouses to get some idea of the types of steel on hand. Do not hesitate to ask for help in locating the structural shapes you need. AR plate is abrasion-resisting plate. Please note that it is a medium to high carbon steel with manganese added. You may experience trouble in welding this material. Your local warehouse can probably have it formed to fit over a standard steel section when wear may be a factor. The bottoms of conveyors are a good place to start.

Cor-ten steel shapes are available now. You might have to ask for HY steels. Check on any low-alloy, high-tensile-strength shapes. The trade names may be the only way they are identified by your supplier. The two main categories of these plates are ASTM A-514 and A-517. Armor Steel produces SSS-100 and 100A and B. Bethlehem Steel has RQ 100-100A and 100B. Great Lakes and Phoenix produce NA XTRA-100 and 110. James and Laughlin Steel has Jalloy S-100 and S-110. U.S. Steel and Leukens both produce T-1, T-1A, and T-1B. T-1 will be mentioned throughout this book, but you should not assume that one company's products are better than another's. The A-517 steels are often called boiler steels, and the A-514 class are structural steels. There is one other steel available in ASTM classification A-203-69. It may not be a true low-alloy steel. The carbon content is low, but in some grades the nickel content may be more than $3\frac{1}{2}$ percent. The other alloys involved are in fractions of 1 percent. Again, this is pressure-vessel steel.

If you have a need for a specific grade, put in an order well in advance. Allow the supplier time to research the types and grades available and ask for all the prefabrication and welding data on a particular steel. The steel companies want the very best results from the use of their products. Follow their recommendations carefully. If preheat and interpass temperatures are needed, simply order color-code crayons from your welding supply house. These crayons work by changing color when a temperature is reached. Using this type of temp-stick and other temperature-testing devices is explained in Chapter 12. These steels are hydrogen-sensitive. Ask your supplier if they have been stored where dew or condensate can form. If you can't be sure, dry the steel with a torch before any welding is attempted.

In choosing the proper steel for the job, don't overlook the fact that a competitor may use more mild steel at a cheaper price.

While we have covered some of the low-alloy steels, we hope that you have become aware that many more steels are available for specific structural use. If you have a need for a property in a construction steel, any good steel warehouse can either deliver or search for a manufacturer that can supply the material. If possible, order by a number.

The shapes available for these steels are not as extensive as what you find with structural grades of mild steel. As you have probably already noticed, most of the fabricated products are welded sections of either flat bar or plate. This is partly because these are heavy construction projects, and standard mild steel shapes present a weight problem. T-1 can be obtained in channel and wide-flange shapes either through a local warehouse or by asking a supplier to order from one of the district centers located to serve all areas in the United States. Angle is not a product sought by fabrication shops, so it is not usually stocked. It could be available as a special order item direct from the factory. Angle should not be fabricated because severe warpage problems will result.

As with mild steel, a sufficient weight of a special order would probably be 62.5 to 69 tons, or approximately 58 to 62 metric tons (rail-shipped in box-car lots). Pricing will be a matter of shipping and storage problems. Since bar and plate are widely used, they will be the lowest-priced items. The lower-tensile-strength products will be cheaper, but the AR plate will cost about the same as T-1. Comparable steels will cost about two and a half to three times the price of mild steel by weight. Always *shop* for the steels you need. Call the local suppliers for direct quotes. If a warehouse has the steel on hand, it may want to sell to restock or change a brand of products. You may find a difference of a few cents per pound or a real bargain! Should you be involved in field erection on a job site far from any supply house, ask for a price delivered where you need the steel. Look out for *FOB*. FOB means *free on board* and, in smaller letters,"our shop." You pay all shipping and handling costs until you sign a delivery invoice at the place where you need the steel. Watch out for prices on remnant steel shapes; it may involve not only the full price, but also a charge for cutting, stocking, or even handling.

Not all prices for materials are add-on prices. Figure 4-1 gives a variety of helpful rolled and formed shapes that will save you time, material, and money. Follow all phone or fax orders with a written request that again shows exactly what you want and what the price quote was on the day you phoned in your order. Suppliers value your business and would not like to lose it due to a misunderstanding in the office.

FARWEST STEEL CORPORATION

TELEPHONE ORDER METHOD
FOR PLATE PATTERN CUTTING

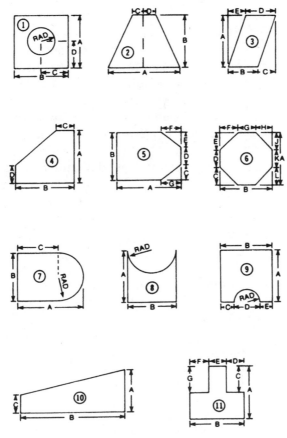

4-1. *"Call" shapes from a full-service steel warehouse.* Farwest Steel

Figure 4-2 shows how to order tapered channel and also conveyor shapes that can be ordered in longer lengths than you may be able to handle in your own shop. It means less handling, fewer joints, and less worry about in-line joining. Figure 4-3 shows plates cut for safety covers of motor drives required by the Occupational Safety and Health Administration (OSHA) for almost all moving-chain or belt-driven machinery. Below these items, you'll notice a sprocket. This company and perhaps your own supplier can deliver such accurate sprockets that roller chain fits close enough to eliminating machining, cutting out pattern-making for cast sprockets. The flame-cutting of wear-resistant steels ensures long life and price breaks that mean savings of 300 to 400 percent over cast materials. The company must charge for these services, but as you can see, they would be cheap at twice the price.

The large warehouses that stock steel products will ship flame-cut T-1 to cambered plate shape ready to be fabricated into special-size beams for low boys, bridges, and other materials that require an arched contour for increased strength. (For a steel arch to fail, sufficient weight must be placed to force the steel to reach a level plain first, and then more weight must be added to begin the rupture of grain structure, causing a crack). Figure 4-4 shows camber or forming of steel shapes so the webs of beams and legs of angles reflect the true shape of contour to match your circular or curved fabrication requirements.

You may feel that since beams can be readily cambered, the price would be less than the fabricated materials. Your engineer may need a depth of sections not available; for instance, a camber beyond mill tolerance or even a flange width not listed for any standard beam. A large warehouse will charge for the entire plate from which the cambered materials were cut, but it will retain the scrap and not usually charge for cutting.

You can now use this information as a guide to what shapes are available, how to use the services of steel warehouses, when it is best to order, how some pre-performed operations are an advantage price-wise, and when and where it may be better to do all the fabrication practices from standard shapes and plates. Price is relative to each project and depends on the complexity of that project and the local availability of the materials needed for completion.

FARWEST STEEL CORPORATION

TELEPHONE ORDER METHOD
FOR FORMING

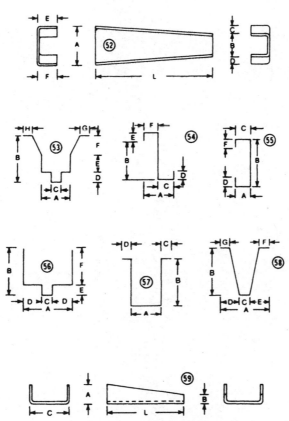

4-2. *How to detail dimensions for formed shapes when ordering from a warehouse.*
Farwest Steel

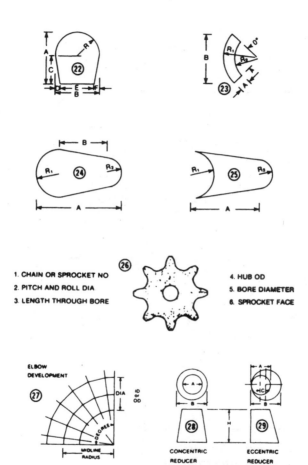

FARWEST STEEL CORPORATION

TELEPHONE ORDER METHOD
FOR PLATE PATTERN CUTTING

1. CHAIN OR SPROCKET NO
2. PITCH AND ROLL DIA
3. LENGTH THROUGH BORE

4. HUB OD
5. BORE DIAMETER
6. SPROCKET FACE

4-3. *Safety and close-tolerance shapes at cost-effective prices.* Farwest Steel

FARWEST STEEL CORPORATION

TELEPHONE ORDER METHOD
FOR ROLLING

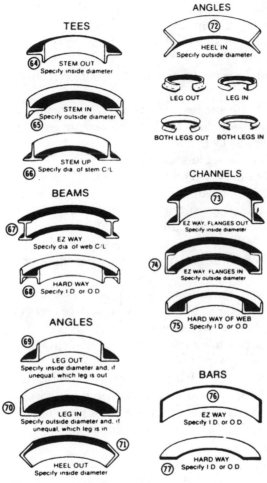

4-4. *The curvature and availability of the shapes.* Farwest Steel

Questions for study

1. Are there reasons for making mild steel your first choice for fabricated products? If so, what are they?
2. Can you list at least three reasons for your choice?
3. Does common sense mean the use of a local supplier only?
4. Can you obtain all structural shapes from your local warehouse?
5. Can you find a reason for fabricating your own structural shapes?
6. Are most of these shapes constructed from channel sections?
7. Would you consider low-alloy, high-strength steels a high cost when shipping is involved?
8. Why not fabricate angle sections?
9. Can you order factory-direct materials?
10. Are standard-weight American tons the same as metric tons?
11. Are all low-alloy, high-strength steels about the same price?
12. Does AR plate cost about the same as T-1?
13. Would you use the telephone to check prices?
14. Will it always be wise to check prices from three or more suppliers?
15. Price-wise, will remnants always be your first choice?
16. Would you sometimes use warehouse services on cut shapes over doing all processes in your own shop?
17. Should you supplement telephone orders with written confirmation?
18. Are cut shapes a better bargain than castings and forgings?
19. What does FOB, our shop, mean to you price-wise?
20. When will your engineer order pieces of plate to be fabricated rather than ready rolled shapes?

5

Cold rolls, alloys, and high carbons

Cold-roll steels, of course, are not all round. Some are, but many are not. The so-called round carbon bars are *drawn* in cold dies in a final finishing process. The carbon content is 1018 to 1020. The last two digits of any American Iron and Steel Institute (AISI) number are the amount of carbon in the steel. This is stated as 0.18 or 0.20 percent, meaning that 0.20 is one-fifth of one percent of carbon. This type of steel cannot be considered even a medium carbon steel.

Round screw stock from which many threaded products are made is AISI B 1112-1113 material and is also a cold-drawn steel. It machines easily and maintains a bright finish. Technically these steels are not structural steel, but they are often used as shafting in fabricated equipment. They are very low tolerance bars, plus 0.0 and minus 0.005 for shafting 4 to 6 inches (101.6 to 152.4 mm) in diameter down to 0.002 for 1 inch (25.4 mm) and under. This is about the same as AISI C 1018-20, but the C tells you that it is turned and polished steel. It must be handled with care—no steel clamping devices, no wire rope, and any welding will undercut the steel at the toes of the weld bead. Cold-finished flat stock is also available but is not widely used. It too, is low carbon steel, C-1020.

Square carbon bars have some use in fabrication, but they are more widely used in machine work. They are stocked in 12-foot (approximately 3.75-meter) lengths. Square and hexagonal bars are close-tolerance products with a 0.004 minus finish.

We must consider cold-drawn seamless tubing to be an advantage in some cases. Mechanical (machined) tubing in steel-rated MT 1015 is often used for the fabrication of cylinders. It is stocked by specialty company warehouses. The yellow pages of the phone directory list tool steel companies. Other cold-drawn tubing is available in both outside diameters and wall thickness.

Tubing with a 10-inch O.D. (outside diameter) has a tolerance of 0.034 inches and, if wall thickness is 0.500 inches or more, it has the same inside tolerance. For general fabrication practices, the cost of these types of steel shapes are out of the question. Although the tubes are available in random length of 17 to 24 feet (approximately 5.20 to 7.20 meters), the cost is given as dollars per inch. Remember that pipe is available in three wall thicknesses and may serve your needs at a fraction of the cost.

The common alloys all have a specific reason for their addition to steel. They also lower the melting point of that steel. In pure form, their own melting point may approach twice that of the alloyed steel. A case in point is tungsten. Metallurgists generally cite its melting point at 4800 to 5400 degrees F (2232 to 2511 degrees C). Steels with a small amount of nickel, chromium, and tungsten have a melting point of only 2780 degrees F (approximately 1297.7 degrees C). The Society of Automotive Engineers (SAE) has adopted AISI standards. The use of alloys in the automotive industry is very widespread and to good purpose.

In large, structural, fabricated items, the cost of alloys is prohibitive, and they are not generally available in common shapes. We have already stated that the last two digits of a AISI number show the carbon content. The first two denote the major alloy or alloys.

Since we are concerned with both fabrication and welding practices, we need not consider any prefix to a number except *H*. Other numbers refer to steel-making processes, while H refers to hardenability factors. The cold forming of these steels is similar to T-1, only in some cases the problems are much greater. The radius allowances must be considered, and it may be necessary to consult the steel maker directly before working the material.

The welding of structural-class alloys will always require use of low-hydrogen or, in some cases, stainless-steel filler metals in the 308 and 309 designations. Butter passing with the previously mentioned stainlesses allows you to attach wear plate or even spring steel to mild steel. I would not under any condition attempt to make a fusion-welded joint of these two steels. Spring steel is high in sulfur content. While sulfur is added to increase machineability in some steels, it produces instant porosity in the welding beads and tends to lead to cracking problems as well.

Some steels containing a small amount of silicon are weldable. Most of the wire filler metals have some silicon in them to improve the wetting action of the weld bead as it is being formed.

When used as alloys, chromium and molybdenum have good impact strength in rolled or drawn tube form. They are used in motorcycle frames and aircraft landing gear. They form well, especially if a small amount of nickel is added. These steels are also used in super-heater tubes in boilers. They are weldable, and chrom-moly filler metals are readily available at a reasonable price. Care must be taken with the welding processes that undercut is not allowed, and all craters must be filled.

Alloys are really specialty products, and their use is often limited to wear-factor conditions or fields where shock, impact, or brittleness would preclude the use of either structural shapes or high carbon steels. In considering common-use alloys, Hadfield manganese is probably at the forefront. It has about 18 percent manganese by volume. It is in widespread use in building and repair of earth-moving and rock-crushing equipment: steam shovels, gas diesel-shovel bucket teeth or the buckets themselves, shovel track pads, rock-crushing rolls (shells), mantles or cone shells for cone crushers, plates for anvil-type crushers, the plates for wear surfaces and perhaps the whole blades and track plates for tractors in the earth moving-industry. At one time, many railroad rails used this steel. Hadfield-grade manganese has a cast hardness factor of 25 on the Rockwell C scale (255 Brinell) as cast or rolled but can harden to 50 Rockwell C (495 Brinell). The impact of rocks actually reduces the crystal size of grain structure in the steel. It presents a very hard surface where wear occurs and a slightly softer but tougher interior.

The space-age alloys of such steels as zirconium and titanium have many uses in the aircraft and aerospace industries. They present very real problems for the welding processes. Some require a vacuum chamber or at least the absence of contaminated atmosphere. The cost factor alone eliminates them from any regular fabrication use.

The use of high carbon steels, cast steels, and cast iron were in widespread use until the low-alloy, high-strength steels came to the market. The high carbon steels are again a relative thing. The percentage of carbon ranges from 0.75 to 1.50. At the high limit, only about 1½ percent of the steel is carbon. The steels that were considered for structural or fabricating use were in the range of AISI 4140 to 4142, which clearly shows that they were medium carbon steels and not truly high carbon.

The first log-loading and -unloading units used to move an entire load of logs rather than one log at a time were partially made of 4140 steel. The bottom two forks, or tines, were cut from one 5-inch-thick plate. Each one was 6 inches wide (approximately 15 cm wide and 13 cm thick), tapering from the 5-inch thickness to a point. The curving top tusks were cut from large plates of 4140 steel 3 to 4 inches (7.7 to 10.2 cm) thick. These tusks ran from 15 inches (38 cm) wide to a point. The scrap material was difficult to use for any purpose. When the use of T-1 steel was introduced into this product, we used boxed sections 1 inch thick by 6 inches wide (approximately 2.6 cm by 15.2 cm). See Fig. 5-1. The top tusks were fabricated from steel ½ inch and ⅜ inch (12.75 to 9.525 mm) thick. The T-1 cost less than one-quarter the price of the 4140, and by fabricating the tusk side plates, there was virtually no scrap. The 4140 had poor impact strength and was subject to time and twisting failure. Due to operator error, one tusk of the box-designed unit was forced completely through a 13-inch (33-cm) log. No adverse damage was noted when the log was cut away.

The parent company that designed and built these machines immediately changed to the use of T-1 fabricated sections. The true high carbon steels are not for fabrication use. They are drill steels, cutlery (knife steels), and many grades of tool steel bits. The tool steel holder bits have up to the following alloy percentages: carbon, 1.5 percent; chromium, 4.75 percent; cobalt, 9 percent; tungsten, 18.5 percent, vanadium, 5 percent; and perhaps molybdenum 1 percent. Such

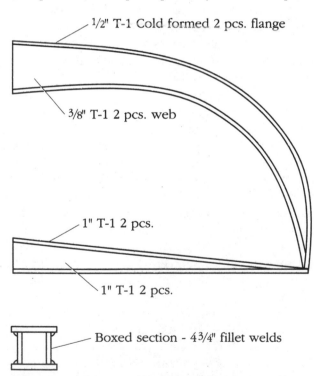

½" T-1 Cold formed 2 pcs. flange

⅜" T-1 2 pcs. web

1" T-1 2 pcs.

1" T-1 2 pcs.

Boxed section - 4¾" fillet welds

5-1. *A detailed drawing of a fabricated item for unloading a truckload of logs at one time.*

steels are often lumped together as tungsten carbide bits. Small pieces of replacement ends are silver-brazed to these holders and can be renewed as they wear down. They are the bits that are used with metal-cutting lathes, boring mills, and steel planes. In a lathe, they will cut steels of 360 to 400 Brinell hardness (42 on the Rockwell C scale). At this time, these are the limits for machining steels.

Although cast iron is not a steel even when alloyed, it can be used as a wear surface when no impact is present. Except for malleable cast iron, it is a very brittle material. The carbon content can range as high as 4 percent, and the Brinell hardness to 555 (Rockwell C to 55).

Questions for study

1. How is cold-roll steel finished?
2. What is the carbon content range for cold-rolled steels?
3. Is cold-rolled steel a high carbon steel?
4. What is the difference between AISI B and C steels?
5. What are the problems in the handling and welding of cold-roll steels?
6. What is meant by MT 1015 material?
7. Can we use MT 1015 for fabrication?
8. How could you find supplies of cold-finished steels?
9. Why would you choose to use pipe rather than cold-finished tubing?
10. What factors would you consider in seeking an alloy steel?
11. How much of manganese is found in Hadfield-grade manganese steel?
12. For what products would you consider the use of manganese steel?
13. What industries might use zirconium and titanium?
14. What are the carbon percentage in AISI 4140 and 4142 steels?
15. Why would you use low-alloy, high-strength steel instead of 4140 steel?
16. Is 4140 a true structural steel?
17. How many real *high carbon* steels are used in fabrication?
18. What is the upper range of carbon content in high carbon steels?
19. How much tungsten can be found in the upper range of a tungsten carbide tool bit?
20. What is the highest range of hardness found in some cast irons?

6

Hand tools

Many of the hand tools used in the fabrication of steel are not pretty. They are of stark utilitarian design.

Measuring tools and a self-chalking chalk line are shown in Fig. 6-1. A self-chalking line works in the same manner as a fishing reel. The line is wound on spool. The top of the case where the retainer ring is attached to the line has a screw top. It can be removed so powdered chalk can be poured into the case. The basic color of the chalk is white. Blue, red, and yellow chalk are also available. Any hardware outlet or welding supply house can supply the chalk. Before this device came into use, the line was wound on a small stick and cake chalk was rubbed on the line for the distance needed. Both items are still available from the same stores. Self-chalking lines like the one shown at the bottom of Fig. 6-1 are much preferred.

To get the best results when producing line marks, follow these directions: Shake the case to ensure good chalk coverage. Pull the line from the case through the hole at the top. Your partner holds one end of the line or case, and you hold the other. Enough line is used to reach the full distance between the points. The line is pulled just fairly taunt. The line is held to the steel at the points used. Reaching out about 2 feet from either end, pull up on the line with your other hand. Raise the line about 6 inches (approximately 15.4 cm) above the steel and then release it. If the line is too tightly held, it will produce one or more phantom lines from the bounce. If it is too loose, it produces a curved line between the points. This line now marked on the steel is fragile. Wind, water, and travel by people will wipe it out. If the line is needed for some time, mark it using a punch.

The measuring tape in Fig. 6-1 has a sturdy metal tape and-crush resistant case. These tapes are generally available only in 50-foot and 100-foot (15.075 and 30.15 meters) lengths. The same company produces tapes running in increments up through 200 feet (60.3 meters). They also produce tapes graduated in feet by $\frac{1}{10}$ and $\frac{1}{100}$. They do make one tape that measures in meters and centimeters. The 25-foot

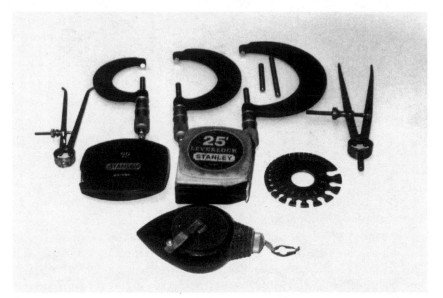

6-1. *Types of measuring tools.* <small>Lane Community College</small>

tape measure in Fig. 6-1 appears larger, but is a heavy-duty tool used by many fabricators. The blade is wider and may feature numbers in larger and contrasting color figures. The longer tapes usually stay in the tool box. The 10-foot, 12-foot, and 25-foot (3.015-meter, 3.61-meter, and 7.6-meter) tapes are always with the fabricator. The person chooses the one that he or she will use most often. These shorter tapes may have a belt or pocket clip for convenience.

The item to the right of the 25-foot tape is a wire gauge. It slips over the edge of a steel sheet or plate to give you thickness measurements. If you usually work with heavy plate, this item is one to carry in your tool box.

One problem with measuring the thickness of sheet or plate is that sheet or plate may have an upset (thickened) edge from shearing. Try the gauge in several places. Your fingers are very good at finding a bulge at a plate edge. Micrometers can reach further from a plate edge and give a more accurate reading. The device at the top left of Fig. 6-1 is a type of inside caliper. It measures hole diameter and distance where other instruments can not reach. The thumbscrew adjustment gives a wide range, and the point-to-point distance is measured by tape, scale, or micrometer.

The three next items are all different-sized micrometers. The company contractor or machine shop will have a set, but the structure steel fabricator will not generally need them. They are precision tools. They are usually calibrated in $\frac{1}{1000}$-inch increments. While machinists may have their own sets up to 6 inches, the company's sets may continue on up to 12 inches (305 mm). The small round rods shown in the curve of the large micrometers are companion anvils, which are used to help the large micrometer measure small items.

The last item in Fig. 6-1 is a type of divider. While it can pick up some measurements, it has better uses. It can be used to lay out small full circles and other curved dimensions. A similar divider, called a wing divider, is a more sturdy counterpart. The legs are heavier and they come in many sizes. If you need to scribe (lay out) many small circles, 12 inches (305 mm) or less on steel, they are excellent. For hard use, you can tip these dividers with hard points. Add a fragment of tungsten carbide or similar tool steel with silver solder. A second method is to add Borrod or stoody #18 to the tips with an oxy-fuel torch. In either case, the tips will last a lifetime. They will add to the depth and clarity of the scribe marks. It will also increase the length of the legs. The marks may be plain enough to follow with a torch.

The first item shown at the bottom of Fig. 6-2 is the big brother of the wing divider. The two trammel points are fixed to any convenient length of metal or wooden stake material. A 1-inch (25.4 mm) standard channel is good for long sweeping radii. A 20-foot (approximately 6.1-meter) length holds its shape well. One trammel device that holds round soapstone crayons can also be bought. These soapstone sticks come in ¼-inch (6.35 mm) rounds and squares. They are also available in ⅛-inch-by-½-inch (3.175-mm-by-12.7-mm) flats. Their uses are covered in Chapter 9.

6-2. *Layout tools.* Lane Community College

The large square with heavy blade shown just above the trammels is really a straightedge. It has a magnet on one side that will hold well on structural steel shapes or plates. A torch can slide along the blade to make a smooth straight cut.

The item to the left and above the magnet square is a small torpedo level. It generally has one glass to show a level flat surface and one to show a vertical 90-degree perpendicular to a flat surface. The large levels may be more accurate in 36-inch and 72-inch (91-cm and 182-cm) lengths.

The tool shown to the right of the level and paralleling the blade of the straightedge is a bevel square. It is used to pick up angles. Its blade length is adjustable, and it can duplicate both acute and obtuse angles. It is a perfect tool for the layout of angles on channel beams and plate. The thumb-screw or wing screw locks the blade position for repeated use. Above the level on the left is a simple protractor. It is not a sturdy tool for measuring angles, but it is not expensive either.

Just above the protractor in Fig. 6-2 is a set of layout tools. At the left end of the blade is a center-head. If the legs touch the sides of any circle, shaft end or pipe, the blade will automatically give you a true diameter line. The tool to its right on the blade is a protractor. It will pick up angles and is durable. Treat it with care and it will last for years. It also features a small level. The last item on the blade is a combination square. It can be reversed on the blade to retain distances between hole centers and from plate and structural shape edges. The front leg also gives a 45-degree angle from the blade. This three-in-one combination tool is excellent for setting saw angles as well as layout applications. The blade is often called a scale. It can be purchased in 6-inch (150.25-mm), 12-inch (305-mm), 16-inch (406-mm), 18-inch (457-mm), and 24-inch (610-mm) blade lengths. The blades are marked in $\frac{1}{32}$-inch or even $\frac{1}{64}$-inch (0.7937-mm or 0.3969-mm) increments. Some tool companies guarantee these items to 0.005-inch tolerances. The blades of all these tools are made of steel.

The long tapes can be clamped to beams to facilitate lengthy layout procedures. The other tools lay on beam or channel shapes. If welding is done on these sections or on sections touching them, the arc may ground through the blades before reaching a ground clamp. This can cut a tape in two pieces or at least ruin it for future use. The heavier blades may have a part removed at the contact point. This is also true if a welding cable has cut insulation and is dragged across the blades. The combination square has been called a tri-square. It is *not*. A tri-square is a solid, nonadjustable square. It is still used by those who work with wood.

The last tool in Fig. 6-2 is a small steel square. It has graduations in
$\frac{1}{16}$-inch (1.5878-mm) increments. Some of the larger squares of this
type have a *body* (blade) 24 inches by 2 inches (610 mm by 51 mm).
The *tongue* (small blade) is 16 inches by 1½ inches (406 by 38 mm).
These larger squares are now available with graduations for each
2 mm. The large framing squares are fine for checking a 90-degree right
angle. The rise and run, angle, and figures are given on some such
squares. They are often used by carpenters for pitch, angles, rafters,
and roofs. If the short leg of these squares were 18 inches (457 mm) instead of 16 inches (406 mm), the 18-inch and 24-inch lengths could
have been divided by 6, resulting in dividends of 3 and 4. By using the
short 3-4-5 method of the Pythagorean theorem, the distance across the
hypotenuse of right triangle would be 30 inches (760.25 mm).

In Fig. 6-3, the tools at the lower left are the dog and wedge combinations. The wedges are tapered, and, when used with the hammers shown in the upper part of Fig. 6-4, they can generate a force of
approximately 50 tons (46 metric tons). The tools to right are center
punches. They are used to indent steel. They can mark hole centers
or any line for permanent reference and also mark lines to be cut
with the torch in adverse visual conditions.

By indenting an area with a punch, you raise the area at its edge.
This method can be used to secure a nut on a bolt. Simply punch the

6-3. *Tools used in the fitting of steel in both the shop and the field.*
Lane Community College

outside pitch of the thread and the nut will not back over the damaged thread area. A shaft can be treated in this manner with a pattern of such indented and raised and center punched areas to provide a temporary means of tightening a loose bearing. This method does not replace lock washers or lock nuts. The shaft will eventually need to be remachined and a sleeve shrunk in place.

If the shaft cannot be replaced, there is an acceptable long-lasting solution. Turn down the shaft to a size .005 of an inch larger than the I.D. (inside diameter) of the tubing needed to fit the bearing. Heat the sleeve to a black red, about 750 to 850 degrees F (285 to 323 degrees C), or very dark red, about 1000 degrees F (380 degrees C). The heating expands the sleeve and it will slip easily over the shaft. When it is in place, the shaft will impart cooling to the sleeve, or you can use water or dry ice to ensure rapid and maximum shrinkage. The punch repair is temporary, but it saves your company down time until the permanent solution is found.

The smaller punch can double as scribe. When it is sharpened to a very fine point, it will be a durable tool for making a scratch line on structural steel. The next tool is a cold chisel. With the hammer shown in Fig. 6-4, it cuts any material not harder than itself, including rivets, bolts, wrought iron, cast iron, steel, brass, copper, and aluminum. Care must be taken when using any type of chisel. The repeated blows from a hammer or pneumatic gun may mushroom (leave ragged edges) on the strike face of the chisel.

Check any type of chisel before use. Grind to original size if conditions permit. Any chisel with a piece missing from the strike face or other part must be discarded. Most of the work previously done by chisels and pneumatic hammers is now done by arc-air metal removal systems.

The next items in line are aligning punches. They extend through one hole or a hole in another plate or shape. They can be left in place to assist the bolting process until other bolts or rivets make them unnecessary. The small dumb bells in the upper right of Fig. 6-3 are aligning tools for open work and are seldom used on fabrication. The set of punches to the right are of different sizes to fit roller chain rivets, etc.

Figure 6-4 shows four types of hammers. The bottom left is a ball peen hammer. It comes in weights from 4 to 40 ounces (11.5 to 117 grams). Often used by machinists, it is also the pipe fitter's friend. It is a good tool for making gaskets for pipe flange seals. The gasket material covers the whole flange face. The flat face of the hammer is tapped completely around the inner diameter of the flange opening, cutting the gasket to exact size. The process is repeated for outside diameter. The hammer head is then reversed. The ball then strikes the bolt holes and cuts the holes in the gasket to size. This is a field operation. Many gaskets are ready-made, and special tools in the shop can make more precise fits.

6-4. *Four types of hammers used for fabrication purposes.* Lane Community College

The hammer on the bottom right of Fig. 6-4 is a cross peen. The head weighs from 2 to 4 pounds (0.9 to 1.88 kilograms). The hammer at the top left is a sledgehammer. These hammers are double-faced and run from 2 to 12 pounds (0.9 to 5.6 kilograms) and are the work-horses for heavy fabrication. The handles are from 14 to 36 inches (38 to 92.5 cm) long.

The hammer at the top right is a personal favorite. The true name may be the hand-drilling hammer. The hammer shown is a Logan Nevada pattern-striking hammer. The handle may be from 10 to 16 inches (26 to 41 cm) in length. In hand-mining today, it is still called a single-jack. The balance can be almost perfect. The handle lengths and grips are suited to an individual. This type of hammer and a star-pointed hand drill made the holes for dynamite charges that produced the tunnels of silver and gold mines in the state of Nevada. Some of the tunnels were more than 9 miles (14.3 kilometers) long.

In buying a hammer for your own use, pick one for balance. Is the handle the right length? Can you swing it easily? Can you stop the swing before an impact point? Someone else might be holding a tool that you must strike at a precise point and angle. The larger the hammer you can handle, the less work you have to do. A blacksmith can tell you that same thing is true of an anvil. A 100-pound (46-kilogram) anvil does half as much work as a 200-pound (92-kilogram) anvil.

The spud wrenches shown in Fig. 6-5 are much-used tools of iron workers and high steel erectors. Most prefer the offset head. It provides greater leverage and, of course, fits only one size nut or bolt head. The tapered handle is a fine tool for aligning bolt holes in matching beams. Its length increases your reach and leverage. Spud wrenches range in length from 12 to 25 inches (approximately 31 to 64 cm). The throat openings are made to fit ½- to 1½-inch (12.7- to 38- mm) bolts. The heads and nut sizes are the important sizes. For erection, ⅝- and ⅞-inch (15.875- and 22.2250-mm) bolts are the usual choices. Most steel erection companies use one of these sizes for every job they do. They can buy the bolts, nuts, and washers in bulk lots. The spud wrenches might also have a quantity price break for a single size.

We cannot do justice to the field of wrenches—there are quite literally thousands of sizes and varieties. Figure 6-6 offers some common examples. The bottom center of the photograph shows a straight box end wrench. The one directly above it is a combination box and open end wrench. The two above that are examples of adjustable open-throat wrenches. If some old-timer asks for a crescent wrench, this is what he or she really needs. The Crescent Tool Company probably produced the first quality adjustable wrench, but there is no data available.

The item above the adjustable wrenches is a speedwrench handle. It fits all sockets for its rated size. If a bolt has lots of thread and you have a deep socket to fit, then this wrench handle could be of value.

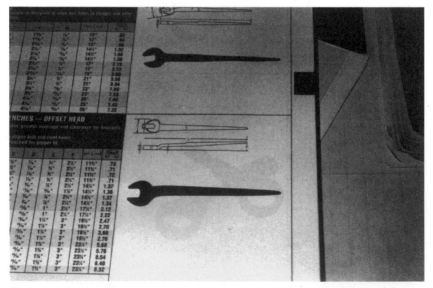

6 5. *Spud wrenches are generally used in steel erection and field fitting procedures.* Lane Community College

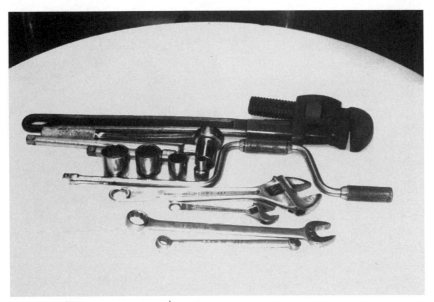

6-6. *A small sampling of the many types and sizes of wrenches available.* Lane Community College

The sockets on the left give you an idea of sizes available. Metric sizes from 6 mm to 41 mm are available at most tool houses. Special sizes can be ordered. There are also universal sockets and deep sockets and a padded deep socket for hard-to-reach spark plugs on engines. Just above the sockets in Fig. 6-6 are two extensions of different lengths. Above them is the ratchet that fits the extensions and the sockets. The advantage of the ratchet is that once the socket fits over the nut, it need not be removed until the operation is complete. The reversible feature means that you can tighten or loosen the nut.

The top tool in Fig. 6-6 is an old-style adjustable pipe wrench. The heavy knurled (deep-grooved) jaws bite into the pipe or fitting for great holding power. They are generally available in sizes to fit pipes from ¾ to 6 inches (19.05 to 152 mm) in diameter.

Welding and cutting equipment

The tools shown in Fig. 6-7 are welding and cutting equipment. The item at lower left is a spark lighter. The industry calls it a striker. In any case, it is the correct tool for lighting any oxy-fuel torch. When the handle is squeezed, the flint scrapes against the ridged steel. The cup just above the flint the place to aim the torch nozzle until ignition occurs.

6-7. *Welding and cutting equipment and safety gear.* Lane Community College

The next item to the right is a set of oxy-fuel tip cleaners. The tiny wires, barely discernible in the picture, fit the holes in the tip. All orifices must be clean. With neutral flame, all fuel gas, short flame points should be the same length. This refers to the cutting head tip shown in Fig. 6-8. When the oxy cutting lever is pressed, the center oxy orifice should produce a light blue narrow column at least 6 inches (155 mm) in length. If this is not the case, you need to clean the offending orifice with the appropriate size wire.

The next tool to the right is a stiff-bristle wire brush. It is used to remove loose scale from steel and loose slag from weld areas. The companion tool to its right is a slag hammer. There are many different patterns of hammer. The blade point on many are turned 90 degrees to the one shown. The handles may be of different materials and shapes. Pick one for the type of weld you need to slag. For manual shielded metal arc welding (SMAW) and *most* heavy slag producing processes, all slag must be removed from each pass and brushed clean before additional welds cover the first pass. Chapter 11 details some processes where the slag can be refloated on the new weld and not include itself in the weld metal.

To the far right of Fig. 6-7 is a pair of welding gloves. They are of heavy leather. Choose a pair that may seem a little large. The air space between your hand and glove will be dead air space insulation. The gloves will also shrink from welding and cutting operations. A welding mitt (not shown) has only a thumb chamber and one large pocket for the rest of the hand.

All steel is hard, sharp, and hot. If you ran into a wood table as a child, you were bruised. Had it been the same thickness of steel, you would have had a cut and perhaps a scar. If you are to work with steel shapes, touch them with your *gloved* hand as you pass to ensure clearance and save on clothing and skin injury. When steel is no longer red, it looks the same as if heat had not been applied. If you handle previously heated steel with a gloved hand, the glove may shrink. If you take the glove off quickly you will have no injury. Note the long gauntlet. It protects the wrist and the material covering your wrist. Short gloves do not protect that area, and if skin is exposed to arc rays it will burn. The longer the exposure, the more damage results. The rays are proven cancer-causing agents.

Do not show other workers how tough you are. The hot metal sparks from a cutting or welding operation can leave small deep pits in the skin. If dirt or grease is left for any length of time in those burns, infection can result.

When exposed to hot metal sparks, you might need to purchase a heavy leather welding jacket or even a complete set of leathers. If you are doing a lot of overhead welding, this is the protection needed.

The unigoggle at the top right in Fig. 6-7 is for cutting or oxy-acetylene welding. It gives a good field of view. The shades of

6-8. *Oxy-acetylene heating or welding and cutting tips.* Lane Community College

protective lenses are stocked by any welding supply store. *Do not* use this equipment for arc welding. As you can imagine, in arc welding, most of your face and hair is exposed to ultraviolet rays and metal spatter. In places where the standard hood will not fit, welding supply warehouse can supply a sock hood. It has a regular viewing lens and cover, but the solid leather hood fits down completely over the head and neck area. Now is a good time to find out if you are claustrophobic. The last item is a type of regular welding hood. There are many types and brand names. All are very safe to use when they are in good condition. Never use a cracked hood or one with a piece chipped from it. Hold the hood to the light; if you see any light, even a pin hole, it is not safe to use. Remember to match the shade of dark lens to the amount of amperage you will be using.

The torches shown in Fig. 6-8 are for cutting, heating, and welding. If you are able to buy only one torch, make it a combination torch. Both heads fit one handle. The heating and welding attachments are the same. A weld supply house can sell you nozzles (tips) from size 000 to size 12. The largest sizes of the heating tips have nozzles that spread the pattern of the orifices. These are referred to as rosebuds. The single-orifice tips are used for welding and direct the heat to one spot only. The cutting tips usually range in size from 000 to 5. The cutting range is from light-gauge sheet metal to slabs 6 inches (155 mm) thick. Since large areas of prefabrication involve out-of-position cutting or one-of-a-kind cutting projects, consider it a full-time job opportunity. If you have the skill, you will be in demand for both shop and field work.

The tools shown in Fig. 6-9 are not strictly hand tools, but you need one or both hands to make them work. The bottom tool is an air-driven power sander-grinder. It should have the same safety shielding of the more recent electrical and compressed air tools. It can be used to remove both metal and slag. No power tool should be used without both face and hand protection. OSHO would rule out the use of this tool until guards were placed around the cutting disk or blade. This is a fine tool for metal preparation.

Do not use stones for preparing aluminum or other soft metal. Some of the abrasive material can remain in the soft metals and cause problems in welding. If the disk-type blade shown is used in deep grooves, care must be taken. The disk may hang up and kick back at the operator.

The hand-drill motor shown above left is electrically powered. Only the chuck size, speed, and amperage rating govern the size drills it will use. It is easy to overload a 3-amp drill. A 5-amp drill or one with higher rating will give you almost continual drilling capacity. These are available in cordless types. Cordless drills are more

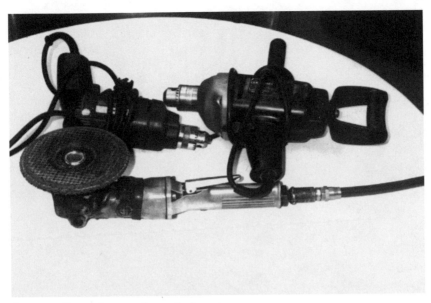

6-9. *Drills and a sander-grinder.* Lane Community College

expensive but very convenient on construction jobs. They are rechargeable and more sturdy than you might expect.

The larger drill on the right has a larger motor and drill size capacity. It can be used by one or two persons. There are many larger two-person drill motors that have a sleeve to fit (Morse taper drills) in place of a key-tightened chuck. Never attempt to operate any drill motor of ³/₄-inch (19.05mm) size by yourself. You and your co-worker should both test the on and off switch. If there is a lock for the on position, test it also. Make sure that the extension cord or power outlet are in easy reach.

The tool shown at the bottom of Fig. 6-10 is a plate clamp. The over-center locking action is positive. The more load that is placed on the clamp, the harder it clamps onto the steel. All of these types of clamps have a safety margin as noted in Chapter 9.

The other tool in Fig. 6-10 is a drill motor attached to a magnetic drill base. It is electromagnetic with a simple on-off switch. It can be locked onto steel in any position. The holding power of the base is usually around 900 pounds for full base contact and does not compute to kilograms per square centimeter. New bases are now available with a narrow oblong shape. This shape fits the faces of smaller beams and channels, allowing full contact of the base. In the round style, if part of the base does not contact steel, the holding power is lessened, and it could pull loose.

Figure 6-11 shows two types of clamps. They both operate on the screw principal. One type of clamp has a protected cover over

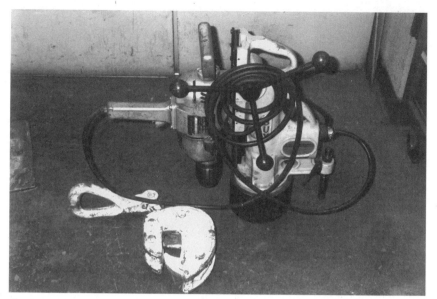

6-10. *A plate clamp (bottom) and a magnetic drill base with drill motor attached (top).* Lane Community College

the screw so that weld spatter does not injure the threads. These short clamps are called C clamps, and this shape reaches over steel parts easily. For as far as they can reach, they will do the work of dogs and wedges.

The bar clamps and pipe clamps simply extend the use of the mechanism. They are often used for aligning or straightening parts in light fabrication situations. An example is shown in Fig. 6-12. Distance A is noticeably shorter than B, and the angles at C and D are obviously more than 90 degrees. After correction clamping, distances A-A and B-B are equal and the parallelogram becomes a true rectangle.

The hand lever hoists in Fig. 6-13 can be used in the same manner on heavy beam fabrication problems. These come in many sizes and are rated as to length of chain, which can be related to lift height and the load weight. The chain type shown is load-capable from ¾ of a ton to 6 tons (0.68 to 5.5 metric tons.) Their lift or pull length is from is from 5 to 20 feet (1.52 to 6.1 meters).

Some tool companies also offer hoists to use the wire rope (aircraft cable) in place of chain. Some heavy-duty hand hoists feature a roller chain. Please note the safety latch on the hooks. Once hooked, no material can slip off no matter how much side or back pressure is applied.

The large item at the bottom of Fig. 6-14 is a turnbuckle. The spud wrench just above it is a construction favorite. The wrench is the handle to run the turnbuckle. It fits in the open slot between the two internal threaded ends. The turnbuckle shown is usually used to tighten wire rope (cable). It can be used to pull beams or even complete structures into alignment. Using one or more on opposite sides will hold a tower or bulkhead in place until it can be welded. Other turnbuckles have clevis ends that can fasten over angle-iron clips or dogs that will act as anchors. The clevis pin is slipped through a hole in the dog or clip. When the dog or clip is welded in place, the turnbuckle is tightened. Using 1-inch (25.4-mm) threaded ends, it will pull or hold with approximately 50 tons (about 45.4 metric tons) of pressure.

The hot slab table shown in Fig. 6-15 is used in conjunction with hand tools. The slabs are about 3 by 6 feet (92 by 184 cm). They are 5 to 6 inches (127 to 154.4 mm) thick. They will support and hold in place tons of steel.

The long-legged hold-down dogs are of manganese steel. To lock onto steel, you simply place as needed and strike the flat area on the top of the dog. To release the piece held by the dog, you strike the other flat place at the back of the dog.

The hold-downs at the upper left of Fig. 6-15 are C clamps with heavy wall square tubing welded to them. The slabs are cast iron and weld spatter does not stick to them.

6-11. *C clamps and bar clamps for aligning steel parts or structures.*

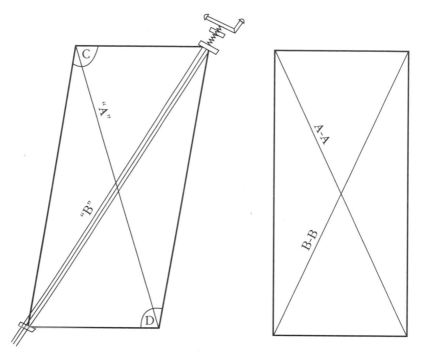

6-12. *Using bar or pony clamps to straighten a part.*

6-13. *A hand-lever hoist, often called a come-along. The gearing gives great mechanical advantage to this hoist.* Western Tool Co.

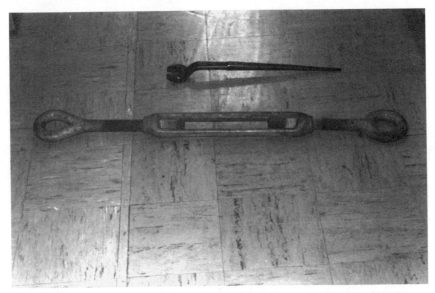

6-14. *A turnbuckle (bottom) and the spud wrench (top) used to move the center body on the threads.* Western Tool Co.

6-15. *A dual-purpose table for layout and forming procedures with two clamping devices.* Lane Community College

Project 5: Square-to-square transition

Using a ³⁄₁₆-inch mild steel plate, lay out and fabricate the transition shown in Fig. 6-16. The material can be sheared to size as soon as you establish the size E-L or length of transition, or K to N, the base of transition. Since ¾ inch equals 1 foot, the distance J-M equals 3 feet. After completing the layout for the true half pattern, use the press brake to finalize each half of the transition. Tack the pieces together to complete the project. You might wish to try straightening out the material in the press for practice plates.

Questions for study

1. What is a chalk line? How many colors of chalk are available?
2. What are the results if a chalked line is stretched too tightly above the steel before it is released?

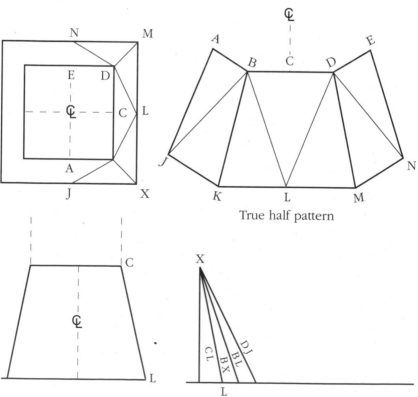

True half pattern

6-16. *Pattern for a square-to-square transition.*

3. How may the lines made by chalk be erased?
4. What length tape would you carry for your own fabrication needs?
5. Name three types of graduated marks available on tape measures.
6. Where are long tapes usually stored?
7. What could you use a wire gauge for?
8. Of what use is an inside caliper?
9. How accurate are micrometers?
10. Why is a wing divider better than the divider shown in Fig. 6-1?
11. Why would you tip wing dividers?
12. When would you use trammel points?
13. How many sizes of soapstone crayons are available?
14. Where would you use a magnetic straightedge?
15. What length levels are available? What are their uses?
16. When would you use a bevel square?
17. Name the two other tools on the blade with the combination square.
18. Can you think of a way to protect tapes and blade-type tools when welding processes are present?
19. Name at least two uses for the common steel square.
20. Are dogs and wedges used more often in the shop or in the field?
21. Name four things that a center punch might be used for.
22. How safe is the mushroomed head of a chisel or punch?
23. In many cases, what has replaced the chisels and hammers?
24. From the descriptions used in the text, could you make a pipe gasket?
25. Name the four hammers shown in Fig. 6-4.
26. What factors would you consider in buying a hammer for your own use?
27. Can you think of three uses for a spud wrench?
28. What is a crescent wrench?
29. What sizes of pipe will adjustable pipe wrenches fit?
30. When and how would you use tip cleaners?
31. Why use gauntlet gloves instead of short gloves for welding and fabrication?
32. What is a sock welding hood?
33. If you can only buy one torch, why choose a combination torch?
34. Why not use grinding stones for preparing brass and aluminum for welding?

35. What shape of magnetic drill base is now available? Why is it better?

36. What is the difference between a bar clamp and a C clamp?

37. How can you tell a rectangle from a parallelogram?

38. What load weights can the hand chain hoists lift?

39. To what height can the chain hoists shown in Fig. 6-13 lift a load?

40. What safety feature is a part of each hook?

7

Power tools

Many of the structural shapes used in building construction require the use of more than manpower. This chapter describes many power tools that are available to the steel fabricator.

Shears (Figs. 7-1 and 7-2) cut material, flat bar, and plate into smaller sections. It is an economical way to prepare parts for the fabrication process. No grinding or shaping is required as for arc or gas cutting.

7-1 *An 8-foot (2.45-meter) shear.* Lane Community College

Brakes (Figs. 7-3 and 7-4) do not brake the material as you might suspect. They form it into shapes and sections ready for fabrication. These sections or shapes would otherwise be made of separate parts and joined by welding or other methods, which would be time consuming and might result in distortion from the joining process.

7-2 *This 12-foot (3.64-meter) shear cuts ¾-inch mild steel.* General Trailer Co.

Some examples of brake-formed parts are noted in Table 7-1. Examine the changes in overall material needed if *inside* or *outside* bends are used. If close tolerances are required, note the thickness of material and the type of bend made. Also, remember that in rolling a

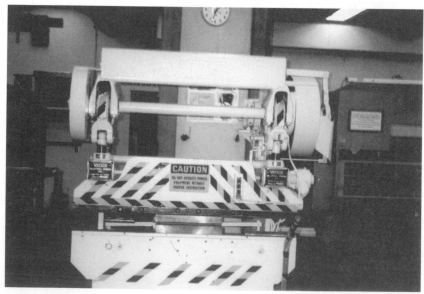

7-3 *A 6-foot capacity power brake.* Lane Community College

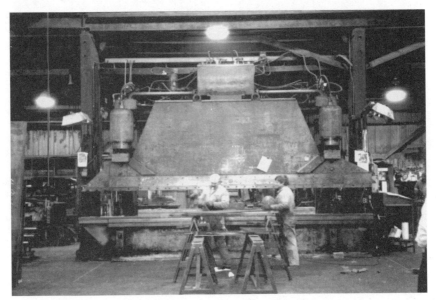

7-4 *This 12-foot (3.64-meter) capacity, hydraulic, motor-driven brake forms ¾-inch (19.05-mm) steel.* General Trailer Co.

cylinder from flat plate, you lose length in direct relation to the thickness of material, resulting in a smaller cylinder than anticipated. A correctly sized cylinder results if the inside diameter required is multiplied by 3.1416 (pi) and the thickness of material is added.

Refer back to Chapter 4 if more explanation is needed in regard to the braking of low-alloy, high-tensile-strength steels or high carbon steels. Remember, the bends must have radiused corners in relation to the thickness of material and the direction the steel was rolled at the steel mill must be known. If the steel was rolled only in one direction, the grain structure will have built in areas of crack-prone material (see Chapter 3).

The "iron worker" (Fig. 7-5) is a trade name for any one of the multipurpose machines made for cutting, punching, and removing small sections of steel. With attachments these machines will work angle, channel, and even small beams. They were originally made for working flat bar, low-alloy steel only, and use an over-center fly wheel.

The shears and brakes use electric motors to drive hydraulic pumps, and, in turn, they drive the pistons of large cylinders for the force necessary to cut and form the steel. Shears and brakes may handle steel to 40 feet in length, but most shops use smaller machines.

Positioners (Fig. 7-6) usually are electric motor-driven and geared to very low speed. They facilitate the welding processes by moving structure into flat and horizontal positions where large-diameter wires

Table 7-1. Calculating the width of plate for bends

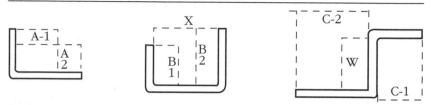

Method for planing plate width when the inside radii of the shape is equal to the plate thickness.

Allow the amount listed to be deducted for inside 90° bends when outside bend dimensions are given.

These are figures for press-brake bends using a V-die or corrected for thickness radius dies.

Thickness given in inches or fractions there of

Figures in column 1 may be multiplied for the number of bends required.

Thickness	1 Bend	2 Bends	3 Bends
$\frac{1}{8}$"	0.20	0.40	0.60
$\frac{1}{4}$"	0.40	0.80	1.20
$\frac{1}{2}$"	0.80	1.60	2.40
$\frac{3}{4}$"	1.20	2.40	3.60
1.0"	1.60	3.20	4.80

Amount to be added to all inside bend dimensions

Thickness	1 Bend	2 Bends	3 Bends
$\frac{1}{8}$"	0.050	0.10	0.15
$\frac{1}{4}$"	0.10	0.20	0.30
$\frac{1}{2}$"	0.20	0.40	0.60
$\frac{3}{4}$"	0.30	0.60	0.90
1"	0.40	0.80	1.20

7-5 *The iron worker cuts, notches, and punches holes in steel.* Lane Community College

and electrodes mean increased deposition rates and better profit margins for companies. These machines are readily available to handle large items to 100 tons in weight. Power-driven rolls are a cheap method of handling cylindrical items. Ball joint positioners that turn

7-6 *The positioner turns a full 360-degree circle and rotates on its axis.* Lane Community College

structures at 360 degrees to either axis lend themselves to manual, automatic, and robotic welding processes.

Cranes

Cranes of various sizes, shapes, and capacities are used in all phases of metal fabrication. Structural shapes and plates arriving from rolling mills and steel warehouses need to be unloaded, stockpiled until needed, moved to fabricated areas, removed from fabricated areas, and loaded for shipment as finished products.

The cranes that are mentioned here are widely used, but there are also many others. They may be referred to as hoists or by colloquialisms such as cherry pickers, gilley-ga-hikes, swing-arms, or others depending on the section of the country where you are employed.

The jib crane was probably first used on ships to stow and unload cargo. The cranes may be small enough to be mounted on the bed of a pickup truck or large enough to handle any type of cargo. Mobile cranes of many sizes and capacities are now used in fabrication and steel construction industries. They may be tire- or track-mounted and can move loads over many types of terrain.

Bridge cranes (Figs. 7-7, 7-8, and 7-9) often span a large expanse (bay) of a fabrication shop. They usually have an electric motor or motors and are run by an operator. The operator may be housed near the top of the lift control area or on the floor of the bay, with hand-held controls. These cranes are often track-mounted and can move loads up and down and to the length and width of the bay.

7-7 *The overhead beam structure of a bridge crane.* Farwest Steel

7-8 *A view of the hoist and trolley mechanisms of a bridge crane.* Farwest
Steel

7-9 *This crane uses an electromagnet, rather than a hook and cable
system, to lift steel.* Farwest Steel

While mobile cranes are often used in the construction of large buildings, they are limited in relation to the height that their booms can reach. The skeletons of all multistoried buildings are fabricated of steel columns and beams. Usually two or more jib cranes are erected at various levels and moved upward as needed.

Gantry cranes, the largest of the breed, are mounted on rails or huge tire-mounted platforms (100 or more tire and axle combinations). The four main support columns that support a turntable mounted boom are massive. Two or more are used to lift a ship into or out of dry dock.

I have not mentioned the major tools for joining structural steels. The bolt-up and riveting guns are almost always driven by compressed air. The bolt sizes are covered by sockets of various sizes and are made to accommodate the drive from $\frac{3}{8}$ to 1 inch. Rivet guns are used in almost the same manner. The sockets fit the rounded head of the rivets, and the backup (bucking) bars or guns have a recessed concave design to match the same size as the rivet head. The rivet heaters are actually small portable furnaces. The heat source is augmented by compressed air, much like the bellows was used to enhance the blacksmith's forge.

The rivet steel allows for rapid expansion and contraction as well as impact when driven by the gun. The heat must be sufficient during the riveting process to allow the second head to form without cracking. The furnace person is as important as the riveter. The rivet must be hot enough to drive on arrival at the point of insertion, but it must not melt around the thin edges of the head while in the furnace. The rivets are picked from the furnace individually with tongs and tossed to the riveter or helper and caught in a funnel, picked out with tongs and placed in the hole for driving. The furnace operator may also have to use a pneumatic tube to send the rivet to the riveter.

As you can imagine, riveting is a highly skilled process requiring complex equipment and competent operators. Metal preparation, such as drilling and reaming holes for close tolerance fit, make this an expensive joining process. However, the joints are more flexible than welds and still have a place in some areas of steel construction.

At this point, I think that cataloging of the types of welding machines would require an entire book. Since many of the machines are dual-purpose and dual-process machines, we must limit this section to tools already discussed. When we reach the section for particular construction fabrication practices, we will attempt to delineate those machines and processes used almost exclusively for those jobs.

The use of automated oxy-fuel cutting equipment has evolved from the track burner shown in Fig. 7-10, which must be hand set as to speed of travel, length of cut, angle of cut or level, and direction of travel, into machines programmed by tape as shown in Fig. 7-11. These newer machines can even be computer operated for the precision cutting of sprockets and other shapes to tolerances that formerly could only be achieved in a machine shop.

7-10 *A simple motorized oxy-fuel cutting machine.* Lane Community College

The use of standard oxy-fuel, multiple-torch cutting equipment, shown in Figs. 7-12 and 7-13, have been a very cost-effective factor for both fabrication shops and field erectors. The best of these machines are available in most large steel warehouse facilities. Since most of the equipment is computer capable, the computer-assisted drawings (CAD) from your drafting department can be sent by modem or diskette for instant use by the cutting department. The quality of the parts cut in this manner is very close tolerance. The types of cutting machines shown in Figs. 7-14, 7-15, and 7-16 can handle steel in thicknesses from $\frac{1}{8}$ inch (3.3175 mm) to 12 inches (30.075 cm).

The materials shown are examples of types of steel cut by Farwest Steel for its customers. These range from the complex shapes pictured in Fig. 7-17 to circles cut from 5-inch- (127-mm-) thick plate (Fig. 7-18). Figure 7-19 shows a contoured plate, laid out and ready for forming in a press brake.

7-11 *A drawing or tape-activated cutting mechanism.* Lane Community College

7-12 *This machine flame-cuts four parts at the same time.* Farwest Steel

7-13 *Computer-operated cutting equipment.* Farwest Steel

7-14 *The table layout shows the scope of material-handling abilities of this machine.* Farwest Steel

7-15 *Dual plasma arc torches.* Farwest Steel

7-16 *This machine is especially suited for thick plate sections.* Lane Community College

7-17 *A variety of precision-cut steel shapes.* Farwest Steel

7-18 *Sample of thick plate cut to a precise diameter.* Farwest Steel

7-19 *The layout lines show this plate is ready for press brake forming.* Farwest Steel

The plasma arc cutting equipment shown in Fig. 7-15 uses the heat of the arc at approximately 10,000 degrees F (5200 degrees C). The use of compressed air rather than costly bottled gases has made this process cost-effective. Its capability of cutting stacked sheets and plates is another plus factor. With the proper use of gases and operator skill, it can be set to cut alloys, stainlesses, aluminum, copper, bronze, and other nonferrous metals. This type of skilled work is usually impossible to do economically in your own shop. The cost of the machines, the training of personnel, and the amount of use should all be taken into account before you purchase a piece of equipment.

The steel-cutting band saw is shown in Fig. 7-20. The capacity of these saws is still increasing and may be limited only by your ability to pay for the size and quality you need. They are available in *dry* or *wet* machines. The wet saws use a water-soluble oil that is directed to the cutting area of the blade. This oil lubricates and cools the steel and the blade during the cutting operation. It is more expensive than the dry saw because it requires a reservoir for fluid, a pump, and straining devices to keep metal filings from clogging the systems. All of these machines require a high degree of operator (fabricator) skill; therefore, you may find that you are stuck as the operator of choice on any one piece of precision power equipment.

7-20 *The band saw is mainly used for straight cuts, but it can be set for miter-degree cutting.* Lane Community College

Forklifts

The forklift (Fig. 7-21) is a power tool that you are probably familiar with. It is easy to operate, and, if battery-run, it offers a pollution-free operation even in a small closed area. It is fine for lifting and moving large beams and completed fabricated sections. It is particularly suited for use in loading or removing materials from trucks or train cars. Since it is available in sizes up to a capacity of 20 tons (18.175 metric tons) and a lift capability up 16 feet (4.95 meters), machines of this type are gas or diesel-operated. Forklifts may be special-ordered with even greater capabilities especially for large construction projects. They would cut down on crane use where both operators and oilers are mandated for some state and federal projects requiring such crane personnel.

The radial-arm drill press (Fig. 7-22) is capable of drilling very thick steel plate. Since the swing arm can cover a 180-degree radius, it is also a fine tool for drilling bolt holes in long beams, channels, and complex combinations of already fabricated shapes. It is a little more difficult to operate than a small press. The chuck to hold small drills can be replaced with a *morse* taper sleeve. The sleeve can handle drills to 3 inches (76.2 mm) in diameter. The use of fly cutters is discussed in Chapter 8.

7-21 *The forklift is the workhorse of many shops, but it is ineffective on soft or uneven ground.* Farwest Steel

7-22 *The radial (swing arm) drill press has lost ground to the magnetic press, especially those that handle variable speeds.* Lane Community College

The pedestal grinders shown in Figs. 7-23 and 7-24 are for heavy-duty metal removal. Any steel that can be brought to their abrasive wheels can be cut or reduced in size. Different types of materials can be used in making the stones. They can have one side for a roughing wheel and the other of finer grit for finish work. Even tool stones are available. Note the guards at the front of the wheels and the safety face shields in Fig. 7-23.

Hoists and air tuggers can use electricity or compressed air as a power source. The air compressors have reserve tanks that are considered pressure vessels. They can hold air pressure rated at 500 psi, or approximately 36.15 kilograms per square centimeter. These tanks are usually licensed and inspected by state or federal employees once a year. You may not alter, repair, or weld on such tanks without current licenses and certifications.

The hoists and tuggers are really nothing more than winch-type drums (holding cylinders for wire rope or cable, with forward, reverse, and stop controls). In conjunction with sheaves and blocks that change the direction and power of the pulling and hoisting operations, they are of extreme value in situations where cranes and other lifting devices do not have ready access. They may have a horsepower rating for the single line from the drum. The controls should be tested and brake mechanisms checked before any load is applied. All wire rope, cable, and auxiliary devices should be checked for damage. One frayed wire is enough to alert you to

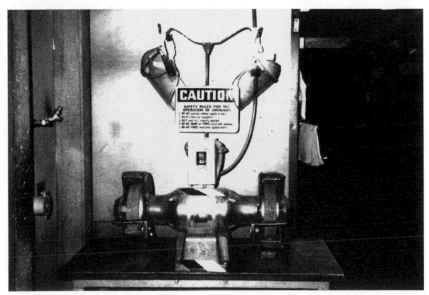

7-23 *Table-mounted dual-wheel grinder.* Lane Community College

7-24 *The grindstones can be replaced by stiff wire wheel brushes on this unit.* Lane Community College

unsafe conditions. Be sure the cable is not kinked or wound incorrectly on the drum.

Compressed air is also used in tungsten arc and carbon arc processes for cutting practices and the removal of metals where oxyfuel processes cannot be used or are not cost-effective. It is also the propellant of choice in the prime coating and painting of structural shapes, finished bridges, tanks, buildings, and fabricated machinery. Follow all safe practice conditions and practices when spray painting.

Questions for study

1. What tools are considered power tools?
2. Does a *shear* cut material?
3. Does a *brake* break steel?
4. Why would you consider bend allowances when calculating size and shape of steel plate for forming processes?
5. When would you want to know the direction or directions of rolling of steel as it came from the mill?
6. What is a positioner and when would you use one?
7. Can you name five types of cranes?
8. What are the largest cranes called?
9. Why not always rivet steel, since it does not affect the grain structure of the material?

10. What is Sunday engineering?
11. What is a track burner?
12. Are there improved oxy-fuel cutting machines available and, if so, what are their capabilities?
13. Why would you sometimes use a metal-cutting band saw over an oxy-fuel process?
14. What kinds of cuts can be obtained from the band-saw cutting process?
15. What is a *wet* saw?
16. What type of motors do forklifts use?
17. How much weight will a large-size standard forklift handle?
18. What is a tugger?
19. Who can alter, repair, or weld on compressed air tanks?
20. Name one factor that might influence your use of a section or coil of wire rope.

8

Working with steel

This chapter looks at the different ways you, as a fabricator, may work with steel. The processes include forming, cambering, cutting, shearing, punching and drilling, alignment, and fitting.

While cold forming has already been discussed in some detail in the preceding chapters, it should also be noted that the forming of large-diameter tubes and pipes can be done with the press brake. It requires real engineering skill and precise data on how much pressure is applied to what thickness of steel and where it is applied. It really might be called a crimping action. Tiny increments on the plate surface and pressure on the V die can be measured in thousandths of an inch and single digits of pressure can be applied. The press brake is also used to preform flat bar on low-boy trailers, as shown previously in Figs. 7-3 and 7-4. The use of low-alloy, high-strength material in thicknesses up to 1 inch (approximately 25.40 millimeters) precludes the use of simple clamping devices to bend the steel. (This would be the equivalent of bending 2½ inches (6.475 centimeters) of mild steel.

The forming of such steels can be accomplished on gradual curving surfaces, as shown in Fig. 5-1 (the log-loading tusks) by the use of portapower or standard hydraulic jacks. (A portapower is a small cylinder head and base, with an hydraulic tube attached to the tank of fluid.) If attachments are not a part of the regular equipment, simply put a cap over the top of the jack. It should be long enough to fit the width of the flat bar and have openings to admit chain links beyond that width. Of course the chain must have a breaking strength equal to the pressure rating of the jack. The jack must operate in all positions; you will be using it upside down. By tack-welding the bar (flange) to the web as they conform to each other, you greatly increase the moment of curve and speed the clamping action. The use of bar clamps may be of help in conjunction with the jacks. I would not recommend this method for bending of these steels beyond ½ inch (12.7 mm) in thickness.

If you are in the field away from all shop devices and you must rough form a curve on plate or flat bar, place the plate or bar across the flanges of a beam or channel section. Then strike the center point with a sledgehammer to indent the steel and put a real curve in steel plate up to ³⁄₈ inch (9.525 mm) thick (Fig. 8-1). It is effective on flat bar up to ³⁄₄ inch (19.050 mm) thick and 8 inches (203.2 mm) wide.

Line of impact points across width of ℝ

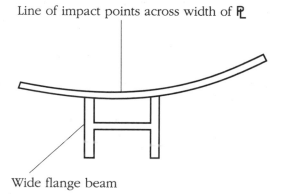

Wide flange beam

8-1 *Rough forming technique where power tools are unavailable.*

The hammer method of prefabrication is also used to form a valley between two adjoining steel structures. A 4-foot- (1.22-meter-) wide plate running the length of the structures allows for drainage from both areas of roof expanse. The plate may be slit at regular intervals to facilitate the bending process. At the plate center, skip 3 inches (approximately 7.62 centimeters) and slit the plate for 12 inches (approximately 30.775 centimeters). After the valley is formed, simply weld the slits.

Hot forming of mild steel plate is quite simple. As previously mentioned, the oxy-fuel torch, small furnace, or forge is used to bring the steel to a dull red color. The steel is now nonmagnetic, and the grain structure has reached a spheroid condition. It bends with its own weight or with little pressure. Even heavy plate 3 to 4 inches thick (approximately 7.65 to 10.19 centimeters) can be bent in a gentle curve if a portion is dogged down and force is applied to the remainder of the plate. Of course, flat bar is formed more easily because it has less mass and the heating is more uniform. The heating and cooling process does not impair the steel in any way.

Cambering can be done using a variety of methods. The first is actually only available on factory order and cannot exceed set tolerances. The formula is ⅛ inch (3.175 mm) × the number of feet divided by 5. Thus 40 feet (12.175 meters) standard (mill beam) multiplied by ⅛ inch (3.175 mm) and divided by 5 gives a maximum camber of 1 foot (30.075 cm). See Fig. 8-2. This is an extreme camber. A more common and practical one would be 3 inches (7.7 cm) for a 40-foot (12.175-meter) length.

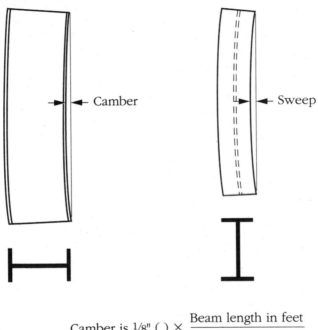

$$\text{Camber is } 1/8" \; (\;) \times \frac{\text{Beam length in feet}}{5}$$

Sweep—due to variations is established in each instance

8-2 *Cambering data on steel mill tolerance for beams.* American Iron and Steel Institute

If a *hot slab* is available (Fig. 6-15), the ends of the section (shape) are dogged down using slab dogs as shown, and the center of the section is pulled or pushed out of line to produce the amount of camber desired. You probably will need to go a little beyond the set figure to allow for shrinkage or spring. Do not heat the steel at only one central point, as this method will result in a kink rather than a uniform curve. The same method could be used without the application of heat, but the spring (attempt to return it to its original shape) would be greater.

The third method is a calculated heating at given intervals along the length of the shape. See Project 2 in Chapter 1. The center V heat uses 6 inches (approximately 15.4 cm) of the top flange and heats all the web as shown to a point at the bottom flange. The heat is 1340 degrees F (approximately 704 degrees C). For the next two areas closest to center, heat approximately 4 inches (10.25 cm) of top flange and about ⅔ of the web to a point. The next two heatings involve 3 inches (7.70 cm) of the top flange to a point about midway in the web. The weight of the beam will act as force. If you want to affect rapid change, more shrinkage, and increased camber, you can apply water to the heated areas. A water spray under pressure is probably best, but rags soaked in cold water serve the same purpose. If more camber is required, you should heat new areas at regular intervals between the previously heated spots and for comparative areas. The amount of camber will be easy to measure in this position by simply stretching a line from end to end and checking the center depth. When you turn the section over, it will show a nice arch.

A fourth method of producing camber is the welding of flat bar to make flanges for a flat plate section that will be the web plate of a beam. The web plate can be cut to almost any curve (camber), and the flange flat bar attached to the web (Fig. 8-3), making sure that they are perpendicular to the web. The welding should be done by two persons. After the tack welds have been made, both welders should start at the same end of the beam on opposite sides of the web. One welder should weld for approximately 3 inches (7.65 cm) before the second welder starts; otherwise, turbulence (arc blow) will produce a rough-looking weld. It will also make welding more difficult for both persons. Welding in this manner will add to the camber because of shrinkage of flange, web, and weld bead material. The weld pull will be about equal and the beam will hold an almost perfect shape.

If little or no more camber is desired than already built into the curved web plate, *backstep* the welds. Backstepping means the laying out of convenient weld distances for the full length of the beam. Weld for a short distance (not more than 2 feet or 0.6175 meters), then skip ahead for an equal distance and weld back until you cover the crater of the first weld. This method breaks up the heat shrink factor and the continuity of heat input. You might also weld a short way, skip to the center, then continue this procedure until the welding is complete.

If additional strength is needed for the beam or beams, you may add flat plate as shown in Fig. 8-4. Not only will the depth of section increase, but the camber will increase and the grain structure will

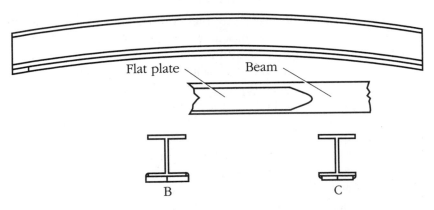

Flat plate Beam

B C

8-3 *Flat bar added to a beam to affect camber.*

shrink all along the heat-affected zones of the welds. Never weld across the flange of a beam in a straight line; it creates a notch effect due to change in thickness of section and *acts* as though you under-cut the beam at the toe of the weld.

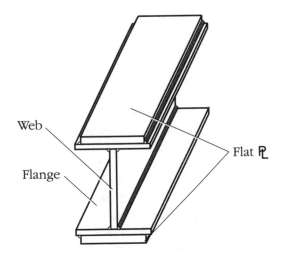

Web

Flat ℞

Flange

Combination section **W-** beam and flate plate fabricated

8-4 *The composite shape when adding flat bar to beams.*

Cutting is usually meant as the oxy-fuel method of both cutting and beveling of any steel shape or section. Some fabrication companies may advertise for help by saying "Cutter Wanted."

The oxy-fuel torch is essential to any fabricator. I have always preferred to use oxy-acetylene. The flame is 6300 degrees F (approximately

3108 degrees C), the hottest of any fuel combination, which means a quicker start, particularly on thick material. As you preheat the steel, watch for the slick appearance in the yellow-orange area. This appearance always tells you that the steel will cut rather than cool off. Since cutting is really a rapid *rusting* process, it is possible that once the *kindling point* has been reached and the steel remains above that temperature, the preheat could be turned off and the oxy alone would continue the cut.

If you must cut laminated material or two thicknesses of steel, slant the torch in the direction of travel and cut the bottom piece first, and the top will take care of itself. The slanting technique is also excellent for sheet steel, because in effect you would be cutting through thicker material, and slag problems will lessen. In both cases, this would probably be a repair or replacement situation. Metal shearing would be a better answer. Laminated or flawed shape or plate should be discarded if possible.

In any case, the fabricator/welder is required to take a welding certification test. Plate tests usually need a backing plate and she or he will have to remove the backing before the testing is completed. The torch is a quick and economical tool for removal.

Figure 8-5 is a good example of layer-by-layer removal of metal without damaging other surfaces. This technique can also be used for cutting out tubes from small stationary boilers without damaging the tube sheets. It is a good way to cut away large nuts that have frozen to studs without damaging the threads on the studs. As you remove one side of the nut, about ⅛ inch at a time, watch for the outline of threads to appear. At this point, stop the cutting process and heat the opposite side of the nut to a dull red color and tap the edges of the nut on the side of the cut away from the exposed thread area. The nut will split open, expanding away from the stud or bolt, and it can be slipped off easily. The torch in a joking way may be referred to as a rubber-handled drill. In a place too tight for a drill to operate, the torch head reaches as needed and can cut or enlarge holes for bolting. Even with practice it is not a close-enough tolerance process for riveting.

If you use either natural gas or propane, note that natural gas has a flame temperature of 4600 degrees F (approximately 2379 degrees C), and propane has a flame temperature of 4579 degrees F (approximately 2295 degrees C). Neither of these gases is suited for welding steel but will do an excellent job for most cutting processes.

Table 8-1 provides information on the cutting of structural steel. It gives correct gas operating pressures, usually lower than most fabrication shops use. Higher pressures mean wasted gas, rough-

8-5 *Layer-by-layer metal removal with an oxy-fuel torch.*

appearing cuts, and perhaps wasted material. You need to know how to cut the beams. Most metal-cutting bandsaws will accept beams to 12 inches in section. While they will cut small beams to any angle, they cannot accommodate the 12 inches and make a 45-degree cut.

If you need to make the cut with an oxy-fuel torch, then you must first lay out the exact angles shown on the print. This requires a protractor and bevel square (both of these tools are shown in Chapter 6). Set the protractor for 45 degrees and set the bevel square from it. Mark the beam with soapstone crayon. Cut the beam as shown in Fig. 8-6. Always start at the bottom of the cut. If you start at the top, the slag will roll ahead of the cut. When the torch reaches the slag, it will not cut but will blow back, causing a rough, slag-filled edge rather than the clean slag-free cut required for welding. Start at A and cut to B, watching the angle needed. If you have some scrap beam, try a practice cut. It may save time and money. When you cut both flanges, you will cut into the web at points C and D. Remove any slag at these points, and reconnect them with soapstone, using the bevel square or any convenient straightedge. When you finish the cut, you will have a beam ready for welding with minimal slag to chip. Most fabricators *never* grind slag. The sander-grinder does not work well in corners, and checking out the equipment and extension cords makes for an inefficient procedure.

Shearing in fabrication shops usually is limited to flat plate or bar stock. The cutting capacity of the machine in Fig. 7-1 is $\frac{3}{8}$ inch

Table 8-1. Gas pressure settings for oxy-acetylene cutting

Metal thickness	Tip size	Cutting oxygen Pressure PSIG min./max.	Flow PSIG* min./max.	Pre-heat oxygen Pressure** SCFH min./max.	Flow SCFH min./max.	Acetylene Pressure** PSIG min./max.	Flow SCFH min./max.	Speed L.P.M. min./max.	KERF width
1/8"	000	20/25	12/14	3/5	3/5	3/5	3/5	28/32	.04
1/4"	00	20/25	22/26	3/5	4/6	3/5	4/6	27/30	.05
3/8"	0	25/30	40/52	3/5	5/9	3/5	5/8	24/28	.06
1/2"	0	30/35	46/58	3/6	7/11	3/5	6/10	20/24	.06
3/4"	1	30/35	70/80	4/7	9/14	3/5	8/13	17/21	.07
1"	2	35/40	110/128	4/9	11/18	3/6	10/16	15/19	.09
1 1/2"	2	40/45	128/140	4/12	13/20	3/7	12/18	13/17	.09
2"	3	40/45	180/200	5/14	15/24	4/9	14/22	12/15	.11
2 1/2"	3	45/50	200/215	5/16	18/29	4/10	16/26	10/13	.11
3"	4	40/50	225/260	6/17	20/33	5/10	18/30	9/12	.12
4"	5	45/55	240/275	7/18	24/37	5/12	22/34	8/11	.15
5"	5	50/55	260/275	7/20	29/41	5/13	26/38	7/9	.15
6"	6	45/55	300/340	10/22	33/48	7/13	30/44	6/8	.18
8"	6	45/55	340/380	10/25	37/55	7/14	34/50	5/6	.19
10"	7	45/55	420/500	15/30	44/62	10/15	40/56	4/5	.34
12"	8	45/55	550/640	20/35	53/68	10/15	48/62	3/5	.41

*This data applicable to 3-hose machine cutting torches only.

**Note: The above data applies to all torches with the following exceptions:

Torch series	Pre-Heat oxygen	Pre-Heat Fuel
MT600 A Series	40 PSIG-UP	2 PSIG-UP
MT500 Series	30 PSIG-UP	5 PSIG-UP

All pressures are measured at the regulator using 25' x 1/4" hose through tip size 5, and 25' X 3/8" hose for tip size 6 and larger. Use 3/8" hose when using tip size 6 or larger.

At no time should the withdrawal rate of an individual acetylene cylinder exceed 1/7" of the cylinder contents per hour. If additional flow capacity is required use an acetylene manifold system of sufficient size to supply the necessary volume.

Victor Co.

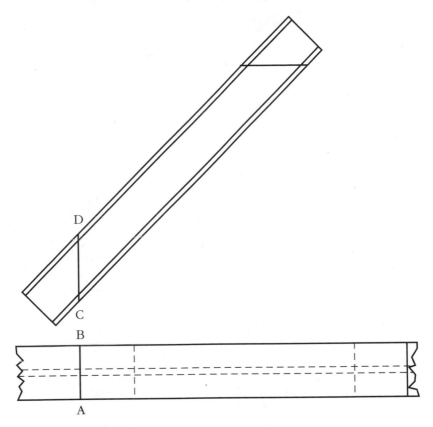

8-6 *Procedures for cutting wide-flange beam.*

(9.525 mm) in thickness and 8 feet (2.45 meters) in width. The shear shown in Fig. 7-2 will cut material ³⁄₄ inch thick and 12 feet (3.6075 meters) wide. While shears do not often cut to machine-finish tolerances, they are all furnished with built-in steel *scales* (measuring devices) calibrated in ¹⁄₁₆-inch (1.5875-mm) increments up to several feet. The shear makes a clean, precise cut, and the steel is ready for fabrication with no more edge preparation unless a bevel is required. Do not shear any material that is not designated as mild steel unless you know that a particular steel is within the rated capacity of your shear. Beware of scrap steel, even in relatively thin sections.

Punching is now done on an iron worker (Fig. 7-5). As mentioned in Chapter 7, it is a multipurpose machine. It is not a precision-process machine. All drill presses and boring mills will work to close tolerances. It is fast, and since most common fabricated items allow for a ¹⁄₁₆-inch (1.5875-mm) tolerance, it is often the tool of choice. The main problems with the process are as follows: Due to the necessary force on the male die, wear is quite rapid. Almost perfect alignment is required for both dies to operate as they should. The *set screws* that lock the dies in place must be checked. Any looseness may let a die slip, and the top die may try to take a fragment out of the bottom die. This could also result in cracking or breaking the top die due to side pressure. Each top die, regardless of size, has a central point. Each time you lay out a hole center, simply use a center punch with a large-size indenter point to enable the punch operator to match it easily to the top die point as the steel slides into position. A true punch press such as that used when multiple parts need to be stamped out requires a skilled operator. Strips of steel usually not over ¼ inch (6.35 mm) thick are hand-fed into dies or a clipper knife. The worker can trip the punch for each operation or hold the switch down and feed the material rapidly enough so that the machine never even pauses. In either case, it produces parts suitable for fabrication purposes.

Drilling can be a precision process, and although holes are often purposely drilled or punched ¹⁄₁₆ inch (1.5875 mm) over size for rapid bolt installation, it need not be the case. If a certain class of fit is required, then all holes will be drilled and perhaps even reamed to exact tolerance. If you want very little play in a bolted joint, simply drill the holes to size and use good-quality cap screws instead of bolts. I will not attempt to give breaking strain figures for the bolt and cap screws, but any industrial fastener company can supply data on size, length, and strength of items. The company will probably fill your needs from stock on hand or can give you an overnight delivery. Drilling can be done using a variety of tools and processes. Small handheld drills can be powered by compressed air, electrical wiring, or even battery use. Larger drill motors to 1 inch (25.4 mm) in capacity require a two-person operation to hold the weight of the tool in place and maintain control of the process.

Many shops and field erection fabricators have chosen to use a magnetic-powered drill base for drilling hard-to-handle shapes or in out-of-position drilling conditions. One person can usually handle the base, turn on the power to the electromagnet locking onto the steel end, then place the drill motor into the clamps that are part of the base mechanism, and you are ready to drill.

Drill *presses* can be used to advantage in all fabrication shops. The simple installing of a drill bit into a key-tightened chuck is much faster than changing dies in an iron worker.

The large radial (swing-arm) drill presses have been in use for a long time. Some date back to a time when water power from a mill race turned a shaft and a belt connected to a stepped pulley (one having different-sized pulleys on a single shaft would give different speeds to the drills).

The radial press shown in Fig. 7-22 handles the cutting of holes up to 5 inches in diameter. Drill bits are readily available in sizes to 2 inches (50.8 mm), but beyond that the cost of each bit is prohibitive unless many holes are required. The alternative method is a cutting operation. A smaller hole is first drilled to fit a shaft, usually 2 inches (50.8 mm) in diameter. The shaft has a hole drilled in it that will take a good-sized tool bit. For a 5-inch (127-mm) hole, a ½-inch (12.7- mm) square tool bit is generally used. The hole in the shaft accepts the bit, and it is locked in place by set screws. This device is called a fly cutter. The speed of the cutter is very slow, measured in revolutions per minute. It peels a shaving as it cuts, so the feed (amount of force downward) is very light. These presses can be set for depth and will self release, so once the process is started, the operator is free to do other work.

Alignment is a function of layout, fabrication, welding, and perhaps straightening. The standard measuring devices all play a part in the process. As you follow the directions from the blueprint, both line items and notes, a pattern should emerge. Each operation will suggest not only what you should do next but how the steel shapes will match each other. The combination square and protractor, along with the framing square, will give you most of the changes in direction for simple fabricated structures. A chalk line is good for snapping lines on steel, but it has a tendency to sag or stretch, particularly when wet or if exposed to wind pressure. Music (piano) wire is a much better choice. Attach the wire to a small turnbuckle and attach the turnbuckle to an upright member, then affix it to another member at a convenient distance and tighten the turnbuckle to stretch the wire. This may be better than a mechanical or laser transit if heavy equipment is operating nearby. These instruments work on a perfect setting of all three legs of a tripod. If one leg moves (is shaken) out of line, all true readings change. Of course, the transit is the correct tool to use if distances are a factor, if elevations are considered for making sure that columns or other structures are in alignment. The transit can also check columns or other sections to show that they are parallel to each other and of equal height.

If you are involved in the shipbuilding industry, you will find that a boat (it is not correctly called a ship until it is commissioned in a formal procedure and named as to class or use) is usually built on a slanted *way*, so when the ship is christened, it can slide down into the water. This presents little or no problem for the shipwrights or shipfitters because a simple device called a declivity board is used in conjunction with a level or plum-bob to show *true* perpendicular alignment of all shapes and sections within the vessel. The declivity board for a specific incline may well be of cast or machine-cut aluminum rather than wood.

The *fairing* of steel plates is a leveling or evening process. It generally involves the use of dogs and wedges. The dog is tack-welded to the lower of the two plates and a wedge is driven into the bite of the dog, which forces the upper plate down while bringing the lower plate up (Fig. 8-7). Fairing may also be accomplished by using a prybar (crowbar). This will work for plates up to $\frac{3}{8}$ inch (9.525 mm) thick. The dog and wedge process will handle much thicker plate. A tapered wedge 1 inch (25.4 mm) to 10 inches (254 mm) in length, when driven with a sledgehammer, will provide approximately 50,000 pounds (41 metric tons) of force. Tack the dog only on the side where the wedge is to be driven for easy removal by striking the other side of dog.

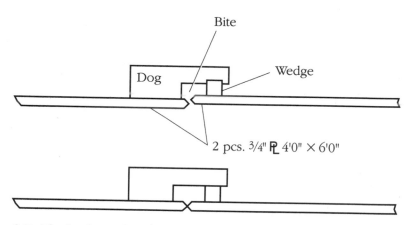

8-7 *The leveling of plate sections.*

Fitting may well have started out as a term for getting a ship ready to sail. It has come to mean any process in the fabrication field where two or more shapes or pieces of steel are positioned ready to be joined, usually by a welding process. The National Union of Boiler Makers, Shipfitters, Blacksmiths, and Welders of America does at least mention fitters. This single union controls most of building and repair of ships within the United States, including naval vessels. Fitters in any other field than shipbuilding are now called fabricators.

Fixtures are those devices that lock pieces of steel into a given completed structure ready for welding. They are often confused with *jigs,* but they usually handle much larger parts or sections and can be turned to any fixed position for welding. Fixtures are almost always hand operated and remain a permanent part of fabrication shop equipment if many individual stock parts are prepared for the parts department for outside sales or delivery.

Positioners come in many shapes and sizes. Some of the very small sizes are used in the making of irrigation pipe fittings. This industry calls them turntables. They are always electrically operated and have a foot control, much like the accelerator of a car, which leaves both hands free for adding fittings and handling the electrode holder or gun. These turntables may have right-angle capabilities.

A set of rolls can also be called a positioner. It most often turns cylindrical parts, such as large-diameter, heavy wall pipe, penstock material, and tanks. If it uses round fixtures in conjunction with track-mounted rolls, it can turn unconventional shapes. One company even employs roll sets for turning small, ocean-going vessels. By now I am sure you realize that the purpose of positioners is to enable welders to use the flat position when welding, which means great savings in time, material, and labor. The welding operators find it less tiring. They can use larger sizes of filler metals without the fatigue of out-of-position work and finish their part of the process and move on to another project. With vertical and overhead welding positions, the loss of bead deposition due to spatter and the limiting factor of how much material will stay on the wall is a terrible cost factor.

Head and tail stock positioners serve almost the same function as the rolls. They simply hold the sections at each end rather than letting them revolve freely. They are much more adaptable to conventional shapes. These machines must operate as one unit. The speed and gearing are factors you must consider to fill your particular needs.

The balanced positioners could well be called unbalanced positioners, except that they use a counterbalance to keep them from falling over when a tack-welded product is attached to the articulated (jointed) arm. In many cases the use of robotics in the welding process has eliminated some of the need for this positioner.

Questions for study

1. Why would you use the press brake to cold-form low-alloy, high-strength steels?
2. Is the resistance of low-alloy, high-strength steels to bending greater than that of mild steel? If so, how much?
3. Name two tools used in the cold-forming of smaller thicknesses of low-alloy, high-strength steels.
4. Can you name two types of heat processes for the hot forming of mild steel?

5. Why does mild steel bend when heated correctly?
6. How badly is the strength of mild steel impaired by the hot bending process?
7. What is camber?
8. Can you give three methods for putting camber in steel shapes?
9. If a compressed-air water spray is not available, what could you use to cool and shrink a steel section?
10. Why not weld across the flange of a beam in a straight line?
11. Can you think of a way to fill a weld crater?
12. What is the *backstep* welding process?
13. What other method could be used to limit continuous heat input?
14. If you advertised for a cutter, what position would you want filled in fabrication shops?
15. How would you remove a nut locked in place by rust from a large stud that you wish to save?
16. Do most fabrication shops use gas pressures that are too low for good cutting practices?
17. With a beam laying on its side, flanges perpendicular to the floor, when cutting with an oxy-fuel torch, do you start at the top or bottom of the flange? Why?
18. Why might you choose a drill press to make a hole rather than punch it with the iron worker?
19. Using a radial-arm drill press, what might you use to cut large-diameter holes?
20. If you need a close-tolerance fit, what might you use in place of bolts?
21. Where would a magnetic drill base be used to drill holes?
22. Can you think of three ways it might be better to use the shear rather than the torch when cutting mild steel plate?
23. Of what use is piano wire in alignment practices?
24. When would a transit be of use in fabrication?
25. How many types of positioners are used?
26. What type of positioner would you use when fabricating penstock sections?
27. Why would you use a positioner?
28. Where is a declivity board used?
29. In what ways are fixtures different from jigs?
30. What is the common name for a small positioner used in welding of irrigation pipe fittings?
31. What is sometimes used to replace the balanced positioner?

9

Shop and safety procedures

The first thing you must do is want to work with those around you and for your company. It doesn't matter if you own a share of the company or are the new kid on the block. If you truly show a willingness to learn and work, you are already a success. Learn to work smart and steadily. The power tools are there for you, and if you don't know how to use them correctly, remember, you won't be the first to ask for help and you won't be the last. Whoever is willing to help you had to first learn from someone else.

One of the best ways to ask for help is to say, "You have the finest skills I have seen. Could you help me to do this job at least well enough so that I don't lose money for our company?" Almost no skilled worker can say no to this approach.

Now we can take a look at the mechanical procedures. Let's start with the layout of the shop. Where does the steel come into a working bay? Will it be stored in racks or stacks? If stockpiled, are the most commonly used materials at the top of the stack for easy access? Is all the steel for a fabrication project stored in one place? Are the plates and sections that require prefabrication work near cutting and forming areas? Are the drawings and blueprints in place? Can you see apparent mistakes on a print? Almost all prints have at least one mistake the first time they are used.

Pay attention to notes on a print; the few minutes spent looking at the print will save a great deal of time later. There will be many things you will need to consider, and they are discussed in detail in this chapter. Will you need to make a floor layout before you move any steel? If plates or beams need to be cut in place with a torch, can you lift the sections with plenty of space to slide scrap steel under the cutting area to protect the floor? The floor must be concrete or steel; do not flame cut over wood. Does it look like the fabricated structure will pass through the doors? What welding processes will you use? How will you move the completed item? Remember, you can not paint in the same area where you use any cutting or welding processes. All steel needs at least a primer coat to resist rust. Some state and federal environmental laws require that a closed area be used so that contaminates from paint do not reach any soil, even indirectly. Check all OSHA rules before building a paint facility. Will you load the finished products on the customer's transport or will you deliver at your expense? Is there a penalty clause in the work contract, and have you completed the job on time?

Organization in this case refers to the personnel involved in a small fabrication shop or plant. If you start out as a one-person shop, you must wear all the hats. You are the owner, stock holder, board member, shop superintendent, foreperson, purchasing agent, equipment operator, prefab worker, fabrication specialist, welder, painter, office boss, phone operator, billing clerk, payroll accountant, tax collector, sales manager, engineer, draftperson, print checker, gopher, and estimator.

The personnel chart in Fig. 9-1 shows the chain of command in a typical steel fabrication plant. The thumbnail sketches may give you a little insight into the workings of a modern corporation. They are intended to encourage thinking about the position you would like to hold and the specific types of knowledge and skills you may need. The lines from one position (or department) to another are flexible and run in both directions. They are intended to show communication and the orderly flow of work progression from one area to another. The board has to establish the production area and scope the financial demands, the labor, materials, resource proximity, and the duration of a product's sales life. It formulates policy on all phases of company operation, preferably in writing on salient points. Factors may dictate courses, but boards must establish direction.

Management follows directives by the board and governs the complex structure known as modern business. It is concerned with many more things than when so-and-so arrives for work, how many welds he or she makes, and when he or she leaves for home. It collects taxes for the government, runs surveys for kindred plants, does research work, supports charities, polices itself, and still finds time to run the business.

The engineer, to you, may be theory and thunder, but by the time you become a foreperson, your salary and his or hers may be more compatible. Your views and terminology will begin to mesh with the engineers. Just think of the engineers as the utility people. They will give you the most efficient method of joining steel. The simple design will have beauty. The cheap method must be the best, and that small beam may be the weight that makes the project completely safe.

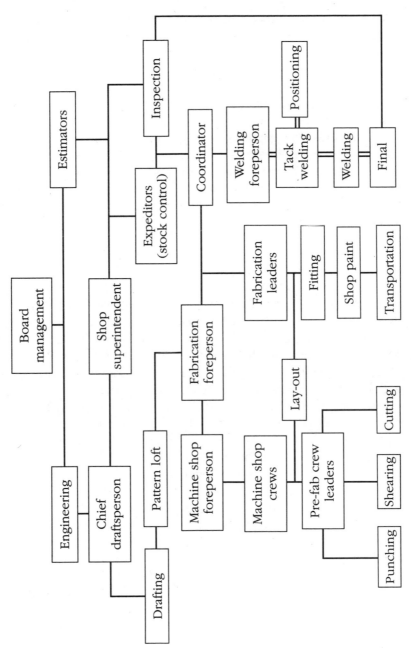

9-1. *A personnel flowchart for a fabrication shop.*

The chief drafter is the in-between person. The sketches, doodles, and figures of the engineer become pictures of production under his or her guidance. He or she works with the engineer to accurately develop the structural conformation that matches the mathematically perfect specifications. The drafting personnel plan and detail it all.

The draftperson works in a clean, well-lighted area, seeking to faithfully reproduce the overall plans, down to each component part. These blueprints will guide each work area and each craftsperson. Back to the drawing board is pretty humorless to these people. At best, an error in dimension may mean lost material. At worst, basic plan changes, details, scraped patterns, forgings, jigs, and materials affect everything and everyone, down to the point of discovery. Bottlenecking operations are in every production department, beyond the point of error.

The fabrication superintendent is usually responsible for all work in progress within the shop.

The estimator makes the company. All large jobs today are the result of bids, whether the customer is a governmental power or the firm next door. The estimator must make an accurate analysis of the cost of the company's production concerning the finished product, allow for a profit margin, and submit the costs for competitive bidding. He or she must calculate the time involved and the cost of completion bid bonding. If the estimator fails to correctly evaluate any one of these factors, he or she stands a good chance of breaking the company. Errors of omission as well as commission must be considered. If the bids are too low, obviously, the company will lose money on the contract. On the other hand, if the bids are too high, the company will receive no contracts and the overhead costs of operation will force closure of the business.

For reasons of brevity, the plant operation chart does not show the business office command progression. It should be realized, however, that the comptroller or bookkeeping department must maintain close dealings with the estimator.

The inspection department must have the good of the company foremost in mind at all times. It should be responsible to no one below the level of shop superintendent. Inspectors must have a thorough knowledge of specifications, such as the types and grades of material and the tests and conditions that products will be subjected to. Then they should be masters in the blueprint and drafting fields. The inspector must check everything from the heat, lot, and grade number of the first shipment of steel to the dimension, shape, and color of the finished product. The quality of workmanship, as it is turned over to the shipping department, is the inspector's final consideration.

The coordinator has only a slightly different job than the expediter. He or she deals with personnel and departments rather than with material. This person must be a diplomat, a yes-person, but most of all, a manipulator. To borrow a sketch from an engineer before it is completed and "mislay" it where it can be found by the chief draftperson can be a tricky business. When the pattern gets a plan before the shop superintendent sees it, the coordinator must have a good alibi. These people will know what has been done and give a kind of approval, but only if it moves the completion dates up the scheduling calendar.

Expediting is a distinct technological field. The foreperson and crew leaders do not have the time necessary to locate and facilitate movement of material through a specific shop area. The expediter must understand production schedules. He or she should check material take-off sheets against production blueprints. This person must know the capacity and, if possible, the operators of all mobile equipment. They nurse the material from the time it is stockpiled until it leaves as a finished product.

The foreperson is responsible. His or her evaluations must consider people even more than products. The capabilities of a worker must be under constant scrutiny. The use of the correct person to do a job is not enough. The upgrading of a craftsperson is a delicate balance from what that person does to what he or she should accomplish for himself or herself and the company. The foreperson must be aware that no one should become indispensable. The foreperson must be available, and his or her view of a problem or a product must be an overall picture. The worker can become involved in some small area, missing an obvious answer or causing a bottleneck, because of lack of perspective.

Crew leaders are subordinate to the foreperson. In small shops, they may be required to work with the tools of the trade. In any event, they must be masters of the craft. It is not absolutely essential that they be well liked, but the level of their technical knowledge and skill must be respected, and their leadership must be backed implicitly by the people in charge.

The skilled worker is the producer. Capital can build nothing by itself. Material wastes away and nonproducers may be eliminated. The worker is worthy of the hire. The energy, skill, and personal pride in the product and company *must* make the company great.

If two or more persons start a business together, we advise against a partnership relationship. If at all possible, have a lawyer draw up corporation papers. The more partners, the more trouble. If any one partner is sued, the other partner or partners are also liable. In a corporate arrangement, only the one sued is liable. The corporation is a separate entity and its assets cannot be touched, nor can the assets of the other stockholders.

Always check with a competent attorney before entering into any business arrangement with other parties. The laws of each state may differ. Please check into the your individual needs and situation before you accept this text as the final word.

The Interstate Commerce Commission (ICC) may require an excise tax payment on materials for transportation across state lines. This may have an upside; if still in effect, an excise exemption certificate may be granted. This requires that materials for such use be sold to you by parts manufacturers or wholesalers at a much lower price. This law was meant to stimulate production of such items on an even basis nationwide. Safety has become a problem for all persons involved in the shop fabrication and erection of steel. An offshoot of safety is product liability. Product liability is much easier to solve. Simply buy some good insurance that will cover the correct or even some incorrect use of your products, materials used, or fabrication practices of your company. The company will probably already have coverage for persons not employed but who may be injured while on company property. Your state probably requires that accident, injury, and death insurance be sufficient to cover all employees. Some companies may be required to cover health insurance. This may include the family as well as those employed. Be informed on such matters before starting work. Other benefits may be the result of union contracts or state and federal negotiations.

Safety rules cover every aspect of this business. In the state of Oregon, the Oregon Occupational Safety and Health code covers a body of administrative rules that are strictly enforced. Violations usually result in *warnings* of unsafe conditions and practices. These must be corrected in a forthright and timely manner. If the matter is not taken care of, fines result.

Remember that safety is always your problem. Common sense is the guide to personal safety. If you feel unsafe in any situation or that unsafe practices are being followed, *stop now*. Bring the problem to the attention of others. If ordered to proceed, think it over; you may need to seek employment with a new company.

Whistle blowers are not popular, but they are still alive. By now, all unsafe practices are covered, and somewhere accident or health reports exist to cover your situation. The following rules apply to the construction and erection practices; they also apply to shops as well. The shops may have other rules governing the use of these materials.

Subdivision H of the Oregon Occupational Safety and Health code covers rigging safety. Rigging covers the loading, unloading, lifting, hoisting, and personnel safety for all operations in the handling of steel. It involves shapes, sections, and fabricated steel both in the shop and at the erection site. Complete tables for lifting loads safely with chains and wire rope cables are available at any state or federal office that administers rules regarding industrial accidents and safety. Check with your local offices for information on safe loads.

OSHA code example

Rigging Equipment For Material Handling

(A) General

> *(1) Rigging equipment for material handling shall be inspected prior to use on each shift and as necessary during its use to ensure that it is safe. Defective rigging equipment shall be removed from service.*

> *(2) Rigging equipment shall not be loaded in excess of its recommended safe working load.*

> *(3) Rigging equipment, when not in use, shall be removed from the immediate work area so as not to present a hazard to employees.*

> *(4) Special custom designed grab hooks, clamps, or other lifting accessories, for such units as modular panels, prefabricated structures and similar materials, shall be marked to indicate the safe working loads and shall be proof-tested prior to use to 125 percent of their rated load.*

(B) Alloy Steel Chains

> *(1) Welded alloy steel chain slings shall have permanently affixed durable identification stating size, grade, rated capacity, and sling manufacturer.*

> *(2) Hooks, rings, oblong links, pear-shaped links, welded or mechanical coupling links, or other attachments, when used with alloy steel chains, shall have a rated capacity at least equal to that of the chain.*

> *(3) Job or shop hooks and link, or makeshift fasteners, formed from bolts, rods, etc., or other such attachments, shall not be used.*

> *(4) Rated capacity (working load limit) for alloy steel chain slings shall conform to safe load values.*

(C) Wire Rope

> *(1) Protruding ends of strands in splices on slings and bridles shall be covered or blunted.*

> *(2) Wire rope shall not be secured by knots, except on haul back lines on scrapers.*

> *(3) The following limitations shall apply to the use of wire rope:*

(I) An eye splice made in any wire rope shall have not less than three full tucks. However, this requirement shall not operate to preclude the use of another form of splice or connection which can be shown to be as efficient and which is not otherwise prohibited.

(II) Except for eye splices in the ends of wires and for endless rope slings, each wire rope used in hoisting or lowering, or in pulling loads, shall consist of one continuous piece without knot or splice.

(III) Eyes in wire rope bridles, slings, or bull wires shall not be formed by wire rope clips or knots.

(IV) Wire rope shall not be used if in any length of eight diameters, the total number of visible broken wires exceeds 10 percent of the total number of wires, or if the rope shows other signs of excessive wear, corrosion, or defect.

General Requirements For Storage

(A) General

(1) All materials stored in tiers shall be stacked, racked, blocked, interlocked, or otherwise secured to prevent sliding, falling or collapse.

(2) Maximum safe load limits of floors within buildings and structures, in pounds per square foot, shall be conspicuously posted in all storage areas, except for floor or slab on grade. Maximum safe loads shall not be exceeded.

(3) Aisles and passageways shall be kept clear to provide for the free and safe movement of material-handling equipment or employees. Such areas shall be kept in good repair.

(4) When a difference in road or working levels exist, means such as ramps, blocking, or grading shall be used to ensure the safe movement of vehicles between the two levels.

(B) Material Storage

(1) Material stored inside buildings under construction shall not be placed within 6 feet of any hoistway or inside floor openings, nor within 10 feet of an exterior wall which does not extend above the top of the material stored.

(2) Each employee required to work on stored material in silos, hoppers, tanks, and similar storage areas shall be equipped with personal fall arrest equipment meeting the requirements of Subpart M of this part.

(3) Noncompatible materials shall be segregated in storage.

(4) Bagged materials shall be stacked by stepping back the layers and cross-keying the bags at least every 10 bags high.

(5) Materials shall not be stored on scaffolds or runways in excess of supplies needed for immediate operations.

(6) Brick stacks shall not be more than 8 feet in height. When a loose brick stack reaches a height of 4 feet, it shall be tapered back 2 inches in every foot of height above the 4-foot level.

(7) When masonry blocks are stacked higher than 6 feet, the stack shall be tapered back one-half block per tier above the 6-foot level.

(8) Lumber

> *(I) Used lumber shall have nails withdrawn before stacking.*
>
> *(II) Lumber shall be stacked on level and solidly supported sills.*
>
> *(III) Lumber shall be so stacked as to be stable and self-supporting.*
>
> *(IV) Lumber piles shall not exceed 20 feet in height provided that lumber to be handled manually shall not be stacked more than 16 feet high.*

(9) Structural steel poles, pipe, bar stock, and other cylindrical materials, unless racked, shall be stacked and blocked so as to prevent spreading or tilting.

TOOLS—HAND AND POWER

(A) Condition of tools: All hand and power tools and similar equipment, whether furnished by the employer or the employees shall be maintained in a safe condition.

(B) Guarding

> *(1) When power-operated tools are designed to accommodate guards, they shall be equipped with such guards when in use.*
>
> *(2) Belts, gears, shafts, pulleys, sprockets, spindles, drums, fly wheels, chains, or other reciprocating, rotating, or moving parts of equipment shall be guarded if such parts are exposed to contact by employees or otherwise create a hazard. Guarding shall meet the requirements as set forth in American National Standards Institute 815.1-1953 (R1958) Safety Code for Mechanical Power-Transmission Apparatus.*

(C) Personal Protective Equipment—Employees using hand and power tools and exposed to the hazard of falling, flying, abrasive, and splashing objects, or gases shall be provided with the particular personal protective equipment necessary to protect them from the hazard. All personal protective equipment shall meet the requirements and be maintained according to Subparts D and E of this part.

(1) All handheld powered platen sanders, grinders with wheels 2 inches in diameter or less, routers, planers, laminate trimmers, nibblers, shears, scroll saws, and jigsaws with blade shanks one fourth of an inch wide or less maybe equipped with only a positive on-off control.

(2) All handheld powered drills, tappers, fastener drivers, horizontal, vertical, and angle grinders with wheels greater than 2 inches in diameter, saber saws, and other similar operating powered tools shall be equipped with a momentary contact on- off control and may have a lock-on control provided that turnoff can be accomplished by a single motion of the same finger or fingers that turn it on.

HAND TOOLS

(A) Employers shall not issue or permit the use of unsafe hand tools.

(B) Wrenches, including adjustable, pipe, end, and socket wrenches, shall not be used when jaws are sprung to the point that slippage occurs.

(C) Impact tools, such as drift pins, wedges, and chisels, shall be kept free of mushroomed heads.

POWER-OPERATED HAND TOOLS

(A) Electric Power-Operated Tools

(1) Electric power operated tools shall either be of the approved double-insulated type or grounded in accordance with Subpart K of this part.

(2) The use of electric cords for hoisting or lowering tools shall not be permitted.

(B) Pneumatic Power Tools

(1) Pneumatic power tools shall be secured to the hose or whip by some positive means to prevent the tool from becoming accidentally disconnected.

(2) Safety clips or retainers shall be securely installed and maintained on pneumatic impact (percussion) tools to prevent attachments from being accidentally expelled.

(3) All pneumatically driven nailers, staplers, and other similar equipment provided with automatic fastener feed, which operate at more than 100 psi pressure at the tool shall have a safety device on the muzzle to prevent the tool from ejecting fasteners, unless the muzzle is in contact with the work surface.

(4) Compressed air shall not be used for cleaning purposes except where reduced to less than 30 psi and then only with effective chip guarding and personal protective equipment which meets the requirements of Subpart E of this part. The 30 psi requirement does not apply for concrete from mill scale and similar cleaning purposes.

(5) The manufacturer's safe operating pressure for hoses, pipes, valves, filters, and other fittings shall not be exceeded.

6) The use of hoses for hoisting or lowering tools shall not be permitted.

(7) All hoses exceeding ½-inch (12.7-mm) inside diameter shall have a safety device at the source of supply or branch line to reduce pressure in case of hose failure.

(8) Airless spray guns of the type which atomize paints and fluids at high pressures (1000 pounds or more per square inch) shall be equipped with automatic or visible manual safety devices which will prevent pulling of the trigger to prevent release of the paint or fluid until the safety device is manually released.

(9) In lieu of the above, a diffuser nut which will prevent high pressure, high velocity release, while the nozzle tip is removed, plus a nozzle tip guard which will prevent the tip from coming into contact with the operator, or other equivalent protection, shall be provided.

(C) FUEL-POWERED TOOLS

(1) All fuel-powered tools shall be stopped while being refueled, serviced, or maintained, and fuel shall be transported, handled, and stored in accordance with Subpart F of this part.

(2) When fuel-powered tools are used in enclosed spaces, the applicable requirements for concentrations of toxic gases and use of personal protective equipment, as outlined in Subpart D, [must be followed.]

(D) HYDRAULIC POWER TOOLS

(1) The fluid used in hydraulic powered tools shall be fire-resistant fluids approved under schedule 30 of the U.S.

Bureau of Mines, Department of the Interior, and shall retain its operating characteristics at the most extreme temperatures to which it will be exposed.

(2) The manufacturer's safe operating pressures for hoses, valves, pipes, fillers and other fittings shall not be exceeded.

ABRASIVE WHEELS AND TOOLS

(A) Power. All grinding machines shall be supplied with sufficient power to maintain the spindle speed at safe levels under all conditions of normal operation.

(B) Guarding. Grinding machines shall be equipped with safety guards in conformance with the requirements of American National Standards Institute, 87.1-1970 Safety Code for the Use, Care, and Protection of Abrasive Wheels, and paragraph (d) of this section.

(C) Use of Abrasive Wheels

(1) Floor-stand and bench-mounted abrasive wheels, used for external grinding, shall be provided with safety guards (protection hoods). The maximum angular exposure of the grinding wheel periphery and sides shall be not more than 90 degrees, except that when work requires contact with the wheel below the horizontal plane of the spindle. Safety guards shall be strong enough to withstand the effect of a bursting wheel.

(2) Floor and bench-mounted grinders shall be provided with work rests which are rigidly supported and readily adjustable. Such work rests shall be kept at a distance not to exceed one-eighth inch from the surface of the wheel.

(3) Cup type wheels used for external grinding shall be protected by either a revolving cup guard or a bent type guard in accordance with the provisions of the American National Standards Institute, B7.1-1970 Safety Code for the Use, Care, and Protection of Abrasive Wheels.

(9) All employees using abrasive wheels shall be protected by eye protection equipment in accordance with the requirements of subpart E of this part, except when adequate eye protection is afforded by eye shields which are permanently attached to the bench or floor stand.

JACKS

(1) The manufacturer's rated capacity shall be legibly marked on all jacks and shall not be exceeded.

(2) All jacks shall have a positive stop to prevent over-travel.

GAS WELDING AND CUTTING
(A) Transporting, moving, and storing compressed gas cylinders
    (1) Valve protection caps shall be in place and secured.
    (2) When cylinders are hoisted, they shall be secured on a cradle, slingboard, or pallet. They shall not be hoisted or transported by means of magnets or choker slings.
    (3) Cylinders shall be moved by tilting and rolling them on their bottom edges. They shall not be intentionally dropped, struck, or permitted to strike each other violently.
    (4) When cylinders are transported by powered vehicles, they shall be secured in a vertical position.
    (5) Valve protection caps shall not be used for lifting cylinders from one vertical position to another. Bars shall not be used under valves or valve protection caps to pry cylinders loose when frozen. Warm, not boiling, water shall be used to thaw cylinders loose.
    (6) Unless cylinders are firmly secured on a special carrier intended for this purpose, regulators shall be removed and valve protection caps put in place before cylinders are moved.
    (7) A suitable cylinder truck, chain, or other steadying device shall be used to keep cylinders from being knocked over while in use.
    (8) When work is finished, when cylinders are empty, or when cylinders are moved at any time, the cylinder valve shall be closed.
    (9) Compressed gas cylinders shall be secured in an upright position at all times except, if necessary, for short periods of time while cylinders are actually being hoisted or carried.
(B) Placing Cylinders
    (1) Cylinders shall be kept far enough away from the actual welding or cutting operation so that sparks, hot slag, or flame will not reach them. When this is impractical, fire resistant shields shall be provided.
    (2) Cylinders shall be placed where they cannot become part of an electrical circuit. Electrodes shall not be struck against a cylinder to strike an arc.
    (3) Fuel gas cylinders shall be placed with valve end up whenever they are in use. They shall not be placed in a

location where they would be subject to open flame, hot metal, or other sources of artificial heat.

(4) Cylinders containing oxygen or acetylene or other fuel gas shall not be taken into confined spaces.

(C) Treatment of Cylinders

(1) Cylinders, whether full or empty, shall not be used as rollers or supports.

(2) No person other than the gas supplier shall attempt to mix gases in a cylinder. No one except the owner of the cylinder or person authorized by him shall refill a cylinder. No one shall use a cylinder's contents for purposes other than those intended by the supplier. All cylinders used shall meet the Department of Transportation requirements published in 49 CFR Part 178, Subpart C, Specification for Cylinders.

(3) No damaged or defective cylinder shall be used.

(D) Use of Fuel Gas—The employer shall thoroughly instruct employees in the safe use of fuel gas, as follows:

(1) Before a regulator to a cylinder valve is connected, the valve shall be opened slightly and closed immediately. [This action is generally termed cracking *and is intended to clear the valve of dust or dirt that might otherwise enter the regulator.] The person cracking the valve shall stand to one side of the outlet, not in front of it. The valve of a fuel gas cylinder shall not be cracked where the gas would reach welding work, sparks, flame, or other possible sources of ignition.*

(2) The cylinder valve shall always be opened slowly to prevent damage to the regulator. For quick closing, valves on fuel gas cylinders shall not be opened more than 1½ turns. When a special wrench is required, it shall be left in position on the stem of the valve while the cylinder is in use so that the fuel gas flow can be shut off quickly in case of an emergency. In the case of manifolded or coupled cylinders, at least one such wrench shall always be available for immediate use. Nothing shall be placed on top of a fuel gas cylinder, when in use, which may damage the safety device or interfere with the quick closing of the valve.

(3) Fuel gas shall not be used from cylinders through torches or other devices which are equipped with shutoff valves without reducing the pressure through a suitable regulator attached to the cylinder valve or manifold.

(4) Before a regulator is removed from a cylinder valve, the cylinder valve shall always be closed and the gas released from the regulator.

(5) If, when the valve on a fuel gas cylinder is opened, there is found to be a leak around the valve stem, the valve shall be closed and the gland nut tightened. If this action does not stop the leak, the use of the cylinder shall be discontinued, and it shall be properly tagged and removed from the work area. In the event that fuel gas should leak from the cylinder valve, rather than from the valve stem, and the gas cannot be shut off, the cylinder shall be properly tagged and removed from the work area. If a regulator attached to a cylinder valve will effectively stop a leak through the valve sear, the cylinder need not be removed from the work area.

(6) If a leak should develop at a fuse plug or other safety device, the cylinder shall be removed from the work area.

(E) Fuel Gas and Oxygen Manifolds

(1) Fuel gas and oxygen manifolds shall bear the name of the substance they contain in letters at least 1-inch high, which shall be either painted on the manifold or on a sign permanently attached to it.

(2) Fuel gas and oxygen manifolds shall be placed in safe, well-ventilated, and accessible locations. They shall not be located within enclosed spaces.

(3) Manifold hose connections, including both ends of the supply hose that lead to the manifold, shall be such that the hose cannot be interchanged between fuel gas and oxygen manifolds and supply header connections. Adapters shall not be used to permit the interchange of hoses. Hose connections shall be kept free of grease and oil.

(4) When not in use, manifold and header hose connections shall be capped.

(5) Nothing shall be placed on top of a manifold, when in use, which will damage the manifold or interfere with the quick closing of the valves.

(F) Hose

(1) Fuel gas hose and oxygen hose shall be easily distinguishable from each other. The contrast may be made by different colors or by surface characteristics readily distinguishable by the sense of touch. Oxygen and fuel gas hoses shall not be interchangeable. A single hose having more than one gas passage shall not be used.

(2) When parallel sections of oxygen end fuel gas hose are taped together, not more than 4 inches out of 12 inches shall be covered by tape.

(3) All hose in use, carrying acetylene, oxygen, natural or manufactured fuel gas, or any gas or substance which may ignite or enter into employee, shall be inspected at the beginning of each working shift. Defective hose shall be removed from service.

(4) Hose which has been subject to flashback, or which shows evidence of severe wear or damage, shall be tested to twice the normal pressure to which it is subject, but in no case less than 300 psi. Defective hose, or hose in doubtful condition, shall not be used.

(5) Hose couplings shall be of the type that cannot be unlocked or disconnected by means of a straight pull without rotary motion.

(6) Boxes used for the storage of gas hose shall be ventilated.

(7) Hoses, cables, and other equipment shall be kept clear of passageways, ladders and stairs.

(G) Torches

(1) Clogged torch tip openings shall be cleaned with suitable cleaning wires, drills or other devices designed for such purpose.

(2) Torches in use shall be inspected at the beginning of each working shift for leaking shutoff valves, hose couplings, and tip connections. Defective torches shall not be used.

(3) Torches shall be lighted by friction lighters or other approved devices, and not by matches or from hot work.

(H) Regulators and Gauges—Oxygen and fuel gas pressure regulators, including their related gauges, shall be in proper working order while in use.

(I) Oil and Grease Hazards—Oxygen cylinders and fittings shall be kept away from oil or grease. Cylinders, cylinder caps and valves, couplings, regulators, hose, and apparatus shall be kept free from oil or greasy substances and shall not be handled with oily hands or gloves. Oxygen shall not be directed at oily surfaces, greasy clothes, or within a fuel oil or other storage tank or vessel.

(J) Additional Rules—For additional details not covered in this subpart, applicable technical portions of American National Standards Institute, 249.1-1967, Safety in Welding and Cutting, shall apply.

(A) Manual Electrode Holders

(1) Only manual electrode holders which are specifically designed for arc welding and cutting, and are of a capacity capable of safely handling the maximum rated current required by the electrodes, shall be used.

(2) Any current-carrying parts passing through the portion of the holder which the arc welder or cutter grips in his hand, and the outer surfaces of the jaws of the holder, shall be fully insulated against the maximum voltage encountered to ground.

(B) Welding Cables and Connectors

(1) All arc welding and cutting cables shall be of the completely insulated, flexible type, capable of handling the maximum current requirements of the work in progress, taking into account the duty cycle under which the arc welder or cutter is working.

(2) Only cable free from repair or splices for a minimum distance of 10 feet from the cable end to which the electrode holder is connected shall be used, except that cables with standard insulted connectors or with splices whose insulating quality is equal to that of the cable are permitted.

(3) When it becomes necessary to connect or splice lengths of cable one to another, substantial insulated connectors of a capacity at least equivalent to that of the cable shall be used. If connections are effected by means of cable lugs, they shall be securely fastened together to give good electrical contact, and the exposed metal parts of the lugs shall be completely insulated.

(4) Cables in need of repair shall not be used. When a cable, other than the cable lead referred to in paragraph (b)(2) of this section, becomes worn to the extent of exposing bare conductors, the portion thus exposed shall be protected by means of rubber and friction tape or other equivalent insulation.

(C) Ground Returns and Machine Grounding

(1) A ground return cable shall have a safe current-carrying capacity equal to or exceeding the specified maximum output capacity of the arc welding or cutting unit which it services. When a single ground return cable services more than one unit, its safe current carrying capacity shall equal or exceed the total specified maximum output capacities of all the units which it services.

(2) Pipelines containing gases or flammable liquids, or conduits containing electrical circuits, shall not be used as a

ground return. For welding on natural gas pipelines, the technical portions of regulations issued by the Department of Transportation, Office of Pipeline Safety, 49 CFR Part 192, Minimum Federal Safety Standards for Gas Pipelines, shall apply.

(3) When a structure or pipeline is employed as a ground return circuit, it shall be determined that the required electrical contact exists at all joints. The generation of an arc, sparks, or heat at any point shall cause rejection of the structures as a ground circuit.

(4) When a structure or pipeline is continuously employed as a ground return circuit, all joints shall be bonded, and periodic inspections shall be conducted to ensure that no condition of electrolysis or fire hazard exists by virtue of such use.

(5) The frames of all arc welding and cutting machines shall be grounded either through a third wire in the cable containing the circuit conductor or through a separate wire which is grounded at the source of the current. Grounding circuits other than by means of the structure shall be checked to ensure that the circuit between the ground and the grounded power conductor has resistance low enough to permit sufficient current to flow to cause the fuse or circuit breaker to interrupt the current.

(6) All ground connections shall be inspected to ensure that they are mechanically strong and electrically adequate for the required current.

(D) Operating Instructions—Employers shall instruct employees in the safe means of arc welding and cutting as follows:

(1) When electrode holders are to be left unattended, the electrodes shall be removed and the holders shall be so placed or protected that they cannot make electrical contact with employees or conducting objects.

(2) Hot electrode holders shall not be dipped in water; to do so may expose the arc welder or cutter to electric shock.

(3) When the arc welder or cutter has occasion to leave his work or to stop work for any appreciable length of time, or when the arc welding or cutting machine is to be moved, the power supply switch to the equipment shall be opened.

(4) Any faulty or defective equipment shall be reported to the supervisor.

(E) Shielding—Whenever practicable, all arc welding and cutting operations shall be shielded by noncombustible or

flameproof screens which will protect employees and other persons working in the vicinity from the direct rays at the arc.

FIRE PREVENTION
(A) When practical, objects to be welded, cut, or heated shall be moved to a designated safe location or, if the objects to be welded cut or heated cannot be readily moved, all movable fire hazards in the vicinity shall be taken to a safe place, or otherwise protected.

(B) If the object to be welded, cut, or heated cannot be moved and if all the fire hazards cannot be removed, positive means shall be taken to confine the heat, sparks, and slag, and to protect the immovable fire hazards from them.

(C) No welding, cutting, or heating shall be done where the application of flammable paints, or the presence of other flammable compounds, or heavy dust concentrations creates a hazard.

(D) Suitable fire extinguishing equipment shall be immediately available in the work area and shall be maintained in a state of readiness for instant use.

(E) When the welding, cutting, or heating operation is such that normal fire prevention precautions are not sufficient, additional personnel shall be assigned to guard against fire while the actual welding, cutting, or heating operation is being performed, and for a sufficient period of time after completion of the work to ensure that no possibility of fire exists. Such personnel shall be instructed as to the specific anticipated fire hazards and how the fire fighting equipment provided is to be used.

(F) When welding, cutting, or heating is performed on walls, floors, and ceilings, since direct penetration of sparks or heat transfer may introduce a fire hazard to an adjacent area, the same precautions shall be taken on the opposite side as are taken on the side on which the welding is being performed.

(G) For the elimination of possible fire in enclosed spaces as a result of gas escaping through leaking or improperly closed torch valves, the gas supply to the torch shall be positively shut off at some point outside the enclosed space whenever the torch is not to be used or whenever the torch is left unattended for a substantial period of time, such as during the lunch period, overnight and at the change of shifts, the torch and hose shall be removed from the confined space. Open end fuel gas and oxygen hoses shall be immediately removed from enclosed

spaces when they are disconnected from the torch or other gas-consuming device.

(H) Except when the contents are being removed or transferred, drums, pails, and other containers which contain or have contained flammable liquids shall be kept closed. Empty containers shall be removed to a safe area apart from hot work operations or open flames.

(I) Drums, containers, or hollow structures which have contained toxic or flammable substances shall, before welding, cutting, or heating is undertaken on them, either be filled with water or thoroughly cleaned of such substances and ventilated and tested. For welding, cutting and heating, on steel pipelines containing natural gas, the pertinent portions of regulations issued by the Department of Transportation, Office of Pipeline Safety, 49 CFR Part 192, Minimum Federal safety standards for Gas Pipelines, shall apply.

(J) Before heat is applied to a drum container, or hollow structure, a vent or opening shall be provided for the release of any built-up pressure during the application of heat.

VENTILATION AND PROTECTION IN WELDING, CUTTING, AND HEATING
(A) Mechanical Ventilation—For purposes of this section, mechanical ventilation shall meet the following requirements:

(1) Mechanical ventilation shall consist of either general mechanical ventilation systems or local exhaust systems.

(2) General mechanical ventilation shall be of sufficient capacity and so arranged as to produce the number of air changes necessary to maintain welding fumes and smoke within safe limits, as defined in Subpart D of this part.

(3) Local exhaust ventilation shall consist of freely movable hoods intended to be placed by the welder or burner as close as practicable to the work. This system shall be of sufficient capacity and so arranged as to remove fumes and smoke at the source and keep the concentration of them in the breathing zone within safe limits as defined in Subpart D of this part.

(4) Contaminated air exhausted from a working space shall be discharged into the open air or otherwise clear of the source of intake air.

(5) All air replacing that withdrawn shall be clean and respirable.

(6) Oxygen shall not be used for ventilation purposes, comfort cooling, blowing dust from clothing, or for cleaning the work area.

(B) Welding, Cutting, and Heating in Confined Spaces

(1) Except as provided in paragraph (B)(2) of this section, either general mechanical or local exhaust ventilation meeting the requirements of paragraph (A) of this section shall be provided whenever welding, cutting, or heating is performed in a confined space.

(2) When sufficient ventilation cannot be obtained without blocking the means of access, employees in the confined space shall be protected by air line respirators in accordance with the requirements of Subpart E of this part, and an employee on the outside of such a confined space shall be assigned to maintain communication with those working within it and to aid them in an emergency.

(C) Welding, Cutting, or Heating of Metals of Toxic Significance

(1) Welding, cutting, or heating in any enclosed spaces involving the metals specified in this subparagraph shall be performed with either general mechanical or local exhaust ventilation meeting the requirements of paragraph (A) of this section.

(i) Zinc-bearing base or filler metals or metals coated with zinc-bearing materials.

Zinc-bearing metal refers to most of the brazing alloys and all galvanized material (pipe and plate). Zinc and all other metals are cumulative poisons when introduced into the body. Iron alone is sloughed off. Zinc in even small amounts produces violent flu-like symptoms. If such sickness does not occur within 24 hours and exposure continues for an extended period, some lesions may appear on the body. The zinc must be scraped from the top of open sores before they can heal.

Zinc chromate is the primer of choice where any steel surface might be impacted by salt or salt spray. In any repair of older fabricated materials (ships, bridges, tanks, and truck frames), the cutting and welding may subject you to exposure and contamination by such primers. The paints themselves will probably contain lead.

Never weld or cut on any tank or container that previously contained volatile fuels such as benzene, diesel, gasoline, or cleaning solvents. Most fire departments have sniffer equipment that can evaluate the explosive danger in most situations. The danger here is that any of the fuels mentioned above can hide in the air spaces in the crystal structure of steel. When heated they come out from between the crystals in the form of vapor. That vapor is highly explosive. Benzene and some cleaning solvents are toxic materials. Benzene is a cancer-causing agent. Care must be taken where fire-retardants containing carbon-tetrachloride may have previously been used. Heat causes them to give off toxic fumes.

> *(ii) Lead base metals*
>
> *(iii) Cadmium-bearing filler materials*
>
> *(iv) Chromium-bearing metals or metals coated with chromium-bearing materials*
>
> *(2) Welding, cutting, or heating in any enclosed spaces involving the metals specified in this sub-paragraph shall be performed with local exhaust ventilation in accordance with the requirements of paragraph (A) of this section, or employees shall be protected by air line respirators in accordance with the requirements of Subpart E of this part.*
>
> *(i) Metals containing lead, other than as an impurity, or metals coated with lead-bearing materials.*
>
> *(ii) Cadmium-bearing or cadmium-coated base metals.*

Many of the silver solders fall into this category. These solders will not be used on any item involved in the manufacturer, preparation, or storage of foodstuffs. Some plain solders contain lead, zinc, and other low-melting-point metals. Radiation safety is not the hazard that you might expect. The highly trained people in the fields have the finest safety record in any comparable industry. When X-rays of steel or welds are being taken, you will be excluded from the area. The whole area will be cordoned off and clearly marked with yellow tape, marked as hazardous. It would be stupid to ignore such signs as these rays can cause sterility and permanent damage to organs of the body.

> *(iv) Beryllium-containing base or filler metals. Because of its high toxicity, work involving beryllium shall be done with both local exhaust ventilation and air line respirators.*
>
> *(3) Employees performing such operations in the open air shall have filter-type respirators in accordance with the requirements of Subpart E of this part, except that employees performing such operations on beryllium-containing base or filter metals shall he protected by air line T aspirators in accordance with the requirements of Subpart E of this part.*
>
> *(4) Other employees exposed to the same atmosphere as the welders or burners shall he protected in the same manner as the welder or burner.*

(D) Inert-Gas Metal-Arc Welding

(1) Since the inert-gas metal-arc welding process involves the production of ultraviolet radiation of intensities of 5 to 30 times that produced during shielded metal-arc welding, the decomposition of chlorinated solvents by ultraviolet rays, and the liberation of toxic fumes and gases, employees shall not be permitted to engage in, or be exposed to, the process until the following special precautions have been taken:

(i) The use of chlorinated solvents shall be kept at least 200 feet, unless shielded, from the exposed arc, and surfaces prepared with chlorinated solvents shall be thoroughly dry before welding is permitted on such surfaces.

(ii) Employees in the area not protected from the arc by screening shall be protected by filter lenses meeting the requirements of Subpart E of this part. When two or more welders are exposed to each other's arc, filter lens goggles of a suitable type, meeting the requirements of Subpart E of this part, shall be worn under welding helmets. Hand shields to protect the welder against flashes and radiant energy shall be used when either the helmet is lifted or the shield is removed.

(iii) Welders and other employees who are exposed to radiation shall be suitably protected so that the skin is covered completely to prevent burns and other damage by ultraviolet rays. Welding helmets and hand shields shall be free of leaks and openings, and free of highly reflective surfaces.

(iv) When inert-gas metal-arc welding is being performed on a stainless steel, the requirements of paragraph (C)(2) of this section shall be met to protect against dangerous concentrations of nitrogen dioxide.

(E) General Welding, Cutting, and Heating

(1) Welding, cutting, and heating, not involving conditions or materials described in paragraph (B), (C), (D) of this section, may normally be done without mechanical ventilation or respiratory protective equipment, but where, because of unusual physical or atmospheric conditions, an unsafe accumulation of contaminants exists, suitable mechanical ventilation or respiratory protective equipment shall be provided.

(2) Employees performing any type of welding, cutting, or heating shall be protected by suitable eye protective equipment in accordance with the requirements of Subpart E of this part.

WELDING, CUTTING, AND HEATING IN WAY OF PRESERVATIVE COATINGS

(A) Before welding, cutting, or heating is commenced on any surface covered by a preservative coating whose flammability is not known, a test shall be made by a competent person to determine its flammability. Preservative coatings shall be considered to be highly flammable when scrapings burn with extreme rapidity.

(B) Precautions shall be taken to prevent ignition of highly flammable hardened preservative coatings. When coatings are determined to be highly flammable they shall be stripped from the area to be heated to prevent ignition.

(C) Protection Against Toxic Preservative Coatings:

(1) In enclosed spaces, all surfaces covered with toxic preservatives shall be stripped of all toxic coatings for a distance of at least 4 inches from the area of heat application, or the employee shall be protected by air line respirators, meeting the requirements of Subpart E of this part.

(2) In the open air, employees shall be protected by a respirator, in accordance with requirements of Subpart E of this part.

(D) The preservative coatings shall be removed a sufficient distance from the area to be heated to ensure that the temperature of the unstripped metal will not be appreciably raised. Artificial cooling of the metal surrounding the heating area may be used to limit the size of the area required to be cleaned. If possible, do not work asbestos in any form!

Each member of any team, department, or group of two or more persons should have a member trained in CPR.

If you suspect an individual has suffered an electrical shock, remove the person from the suspected source using *dry fiber rope* or *dry wood*. Once the person is free, start life-saving techniques immediately. The first 10 seconds are critical.

When guiding a load of steel being moved by a crane, do not touch the steel directly. Use a fiber rope to steer the material.

If you are pressed into service as a rigger, you shall be trained in all hand signals before attempting to communicate directions to a crane operator. If land lines (telephone or walkie-talkies) are to be used, question the crane or hoist operator to make sure directions are clear. Only one person may direct an operator. If only your head and shoulders are above ground level, you are considered below ground level. You can never leave one person alone below ground level. If injured, you may not be noticed.

When operating power equipment, especially that which is blade-equipped or has continuously moving parts, two or more persons must be present.

All persons should attend regularly scheduled safety meetings. Before attending, write down at least one item that should be considered. Since injury or wrongful death law suits can bankrupt almost any company, you can be sure that even things you might think to be trivial will receive prompt attention.

Drawings and sketches usually refer to minor operations in the fabrication industry, perhaps a one-time small change in steel shape or sections, a method of preparing parts, or welding and cutting procedures. Drawings, however, must contain enough detail to enable the fabricator/welder to complete the project in an acceptable manner with good appearance. Always ask for (demand) at least a sketch to protect yourself in case of changes made at a higher level. Sketches should contain all dimensions, types of steel to be used, how this part fits with existing structures, and, if possible, notes on prefab and welding processes.

Blueprints show single-dimensional views of three-dimensional objects or structures. Once you master the art of looking at two or more of the views and can envision the whole object or fabricated structure, you will quickly become the fabricator that all companies want. All three views have names. The first is a *top view* (looking only at the top surface). The second is a *front view*, looking only at the front surface, and usually the third looks only at the *right end surface*. If there is a special need for showing it, you must clearly label a left end view. In extreme cases, you might be shown a back view or a bottom view. These last three views will only be used if some important feature of the item can be shown in no other way. While a draft person or a CAD program will label the views in this manner, a trained engineer will call them a *plan view*, a *front elevation*, and an *end elevation*.

The first skill you will need is to identify all lines on the print. Refer to Fig. 9-2. Solid dark lines show outlines that are continuous. The intermittent lines are hidden lines, meaning that they indicate shapes behind the face of a given view. If you are looking at a steel plate, you do not see the steel beams in back of the plate that act as support members. The beams must be shown in another view.

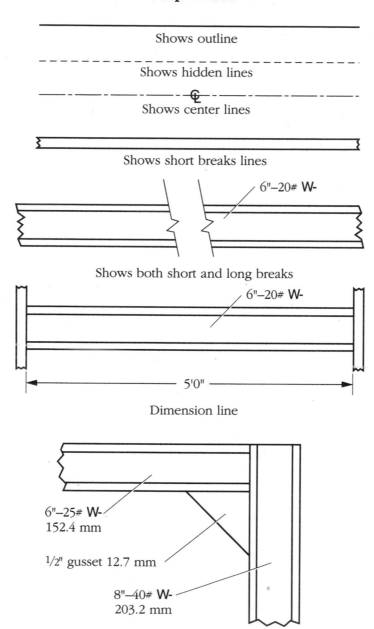

Shows outline

Shows hidden lines

Shows center lines

Shows short breaks lines

6"–20# W-

Shows both short and long breaks

6"–20# W-

5'0"

Dimension line

6"–25# W-
152.4 mm

1/2" gusset 12.7 mm

8"–40# W-
203.2 mm

Shows a reference line

9-2. *The meaning of the lines shown on blueprint.*

Center lines should always be identified by a C superimposed on an L. If the draftperson fails to clearly mark a center line with the C and L brand, he or she might show it as three short lines and then one longer line that continue for the length of the item to be fabricated.

Short irregular lines at the end of shape or section show a break in the print, not in the steel itself. The print maker did not have room on the print to show the other 13 feet that are of exactly the same steel shape or complete structure. A pair of long black lines with very regular lighting strikes or reverse Z configurations separated by a blank space indicates a long break. A dimension line begins and ends with arrow points; its length as measured between the arrow points is centered in the middle of the line.

A reference line has one arrow point, and the point is directed to a shape or item. The other end of that line tells you exactly what the shape or item is. The weld symbol is truly given on a reference line. It and its component parts are outlined later in this chapter.

While this text will show several blueprints, it is not intended to replace a course in blueprint reading. Figures 9-3, 9-4, and 9-5 should be of some help in the study of prints from which a fabricator will have to work.

The use of special materials and processes may be more easily understood on blueprints by the use of notes and specifications. Notes and specification may be placed near the material, the view using the process, and in the specification area shown in Fig. 9-6. If we had to detail such information, blueprints would be unwieldy, needing pages of written work.

A note may be printed in full, abbreviated, or mathematically identified as a dimension. It may be clearly shown by the use of a leader line or at least placed adjacent to the thing it refers to if it is not a general note.

A *general note* is put in a conspicuous place (usually the right side of the print) and gives a more detailed explanation. It is set apart from any one view. A general note might be shown as follows on Fig. 9-7:

Note: THE LAST TWO SECTIONS OF THIS TOWER WILL BE BOLTED ONLY. FOLLOW ERECTION BOLT SPECIFICATIONS ASME #10.

We will not attempt to give a complete course in blueprint reading. It is to your advantage to take as many advanced courses in this field as are available to you. They might be aimed at rapid familiarity of complex fabricated structures, recognition of real dimensions from scaled drawings and figuring weights of shapes and sections. Then the next step might be a course in estimating.

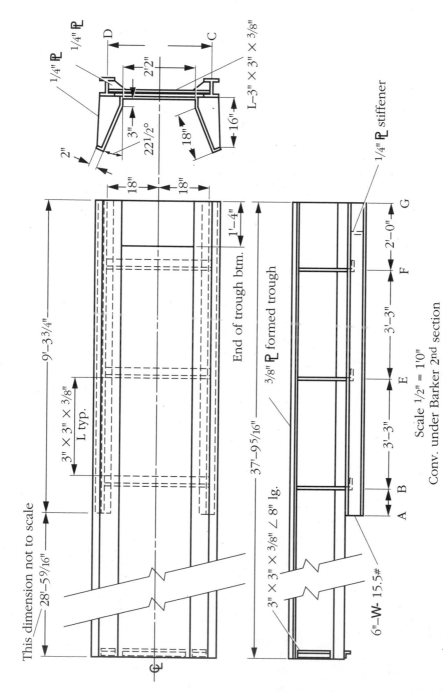

9-3. *A blueprint detailing the fabrication of a conveyor.*

The bidding for fabrication contracts requires absolute accuracy. If a bid is too high, some other company will get the job. If too low, your company will lose money. If your company is small or a bid is required instantly, you can use a simple rule of thumb. Calculate the pounds of steel needed for the job. Call one steel supply company for a price per pound. Now, double that figure and add 10 percent. The doubling figure is a ball-park estimate for the shop labor cost. The 10 percent is for business costs and other contingencies.

For example, if a job calls for 4000 pounds (2 tons or approximately 1.8125 metric tons) at 36 cents per pound, your cost would be $1440. Double that figure to $2880 and add 10 percent, or $288. The correct quote would be $3168. This method will not work for a project involving many small parts or machine work or outside labor, but it works well for mild steel plate, angles, channels, and beams only. Do not use this method for large, complex fabrication projects.

If a large corporation would even accept the small job noted in the example, the cost of capital outlay and paying many nonproduction but necessary persons would usually result in lost dollars. A small shop with eight or fewer employees can always compete with much larger

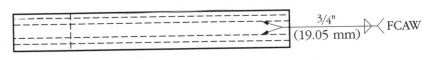

Plan view

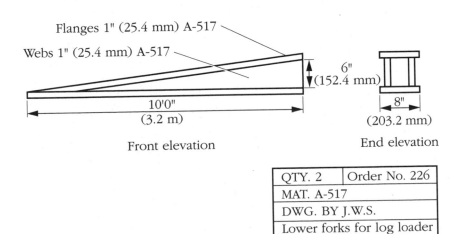

9-4. _Blueprint of lower fork of log-loading machine._

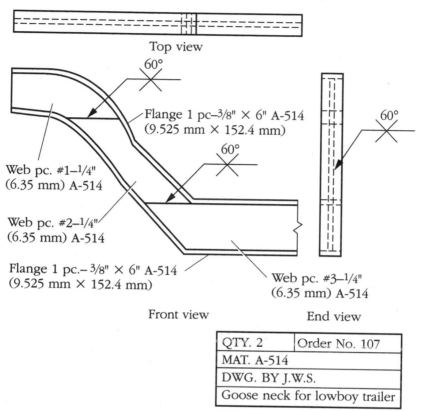

9-5. *Blueprint of gooseneck section of a lowboy trailer.*

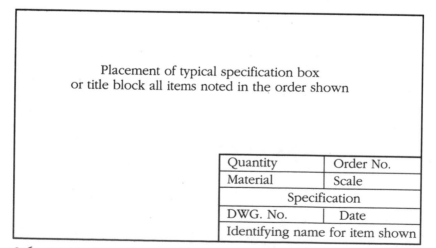

9-6. *A typical specification block or box on a blueprint.*

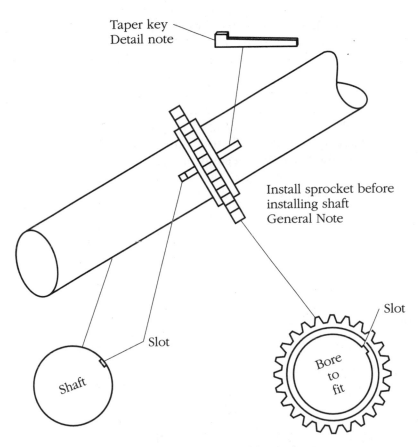

9-7. *Placement of a general note on a blueprint.*

corporations. The number of eight employees was not picked from the air but is calculated from many costly studies on effective management processes. One person must manage the small shop, the workers, production schedules, equipment usage, and consumable supplies inventory. Another must handle sales, bids, making blueprints, simple engineering from proven tables, and purchasing. Both of these people must also be excellent craftspersons. They must be able to do their many jobs and still spend a few production hours each day.

Symbols

Symbols must be used to cut down on written or printed notes shown on prints and drawings. Symbols are also used to show processes and points for nondestructive testing. Figure 9-8 shows the welding symbol. The American Welding Society simplified and consolidated all the components necessary for detailing welds on blueprints. It may seem complicated and cumbersome, but with a little study, the weld symbols on prints can be easily understood. They show you where details of all welding references should appear on the blueprint arrow reference line. Figure 9-9 clarifies the placement of welds. Arrow, side, and other side references might confuse the new fabricator/welder.

Figure 9-10 shows both old and new field weld symbols. The round black circle was originally used and is now not used except by those who learned in a drafting study many years ago. The flag comes more quickly to your attention. The previous symbol was often mistaken for the weld all around symbol.

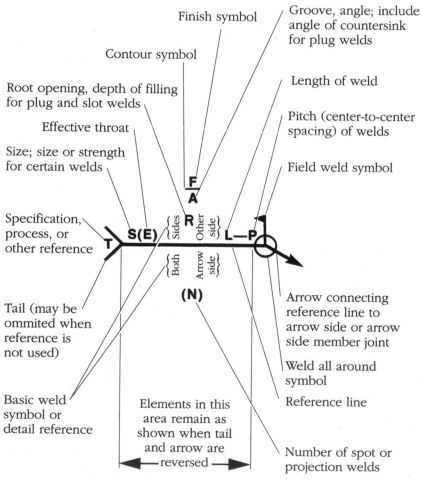

Finish symbol

Groove, angle; include angle of countersink for plug welds

Contour symbol

Root opening, depth of filling for plug and slot welds

Length of weld

Effective throat

Pitch (center-to-center spacing) of welds

Size; size or strength for certain welds

Field weld symbol

$\dfrac{F}{A}$

Specification, process, or other reference

S(E)   R   L—P

T   Sides   Other side

Both   Arrow side

Arrow connecting reference line to arrow side or arrow side member joint

(N)

Tail (may be ommited when reference is not used)

Weld all around symbol

Basic weld symbol or detail reference

Elements in this area remain as shown when tail and arrow are ◄— reversed —►

Reference line

Number of spot or projection welds

9-8. *A single symbol that covers all welding data.* American Welding Society

BASIC WELDING SYMBOLS AND THEIR LOCATION SIGNIFICANCE

LOCATION SIGNIFICANCE	SQUARE	V	BEVEL	GROOVE U	J	FLARE-V	FLARE-BEVEL
ARROW SIDE	(symbol)	(symbol)	(symbol)	(symbol)	(symbol)	(symbol)	(symbol)
OTHER SIDE	(symbol)	(symbol)	(symbol)	(symbol)	(symbol)	(symbol)	(symbol)
BOTH SIDES	(symbol)	(symbol)	(symbol)	(symbol)	(symbol)	(symbol)	(symbol)
NO ARROW SIDE OR OTHER SIDE SIGNIFICANCE	NOT USED	(symbol)	NOT USED	NOT USED	NOT USED	NOT USED	NOT USED

SUPPLEMENTARY SYMBOLS

WELD-ALL-AROUND	FIELD WELD	MELT-THROUGH
(symbol)	(symbol)	(symbol)

CONTOUR		
FLUSH	CONVEX	CONCAVE
(symbol)	(symbol)	(symbol)

BASIC WELDING SYMBOLS AND THEIR LOCATION SIGNIFICANCE

LOCATION SIGNIFICANCE	FILLET	PLUG OR SLOT	SPOT OR PROJECTION	SEAM	BACK OR BACKING	SURFACING	FLANGE EDGE	FLANGE CORNER
ARROW SIDE	(symbol)	(symbol)	(symbol)	(symbol)	(symbol)	(symbol)	(symbol)	(symbol)
OTHER SIDE	(symbol)	(symbol)	(symbol)	(symbol)	GROOVE WELD SYMBOL	NOT USED	(symbol)	(symbol)
BOTH SIDES	(symbol)	NOT USED	NOT USED	NOT USED	NOT USED	NOT USED	NOT USED	NOT USED
NO ARROW SIDE OR OTHER SIDE SIGNIFICANCE	NOT USED	NOT USED	(symbol)	(symbol)	NOT USED	NOT USED	NOT USED	NOT USED

9-9. *The weld symbols used on a blueprint.* American Welding Society

Supplementary welding symbols

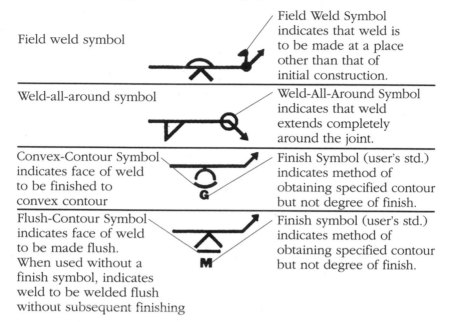

Field weld symbol	Field Weld Symbol indicates that weld is to be made at a place other than that of initial construction.
Weld-all-around symbol	Weld-All-Around Symbol indicates that weld extends completely around the joint.
Convex-Contour Symbol indicates face of weld to be finished to convex contour	Finish Symbol (user's std.) indicates method of obtaining specified contour but not degree of finish.
Flush-Contour Symbol indicates face of weld to be made flush. When used without a finish symbol, indicates weld to be welded flush without subsequent finishing	Finish symbol (user's std.) indicates method of obtaining specified contour but not degree of finish.

9-10. *Both old and new field weld symbols.* American Welding Society

The plug and slot welds are not used as often as they should be. They cause little or no distortion and make the fitting of one shape to another without beveling for weld accommodation at edges that would otherwise have to be welded.

The symbols mentioned for nondestructive testing are not widely used and are clearly stated if required by particular or general notes. The symbols may not be used, but the tests themselves may be required in a stated code or as a contractual obligation. The use of such tests will be spelled out in detail. These symbols are only mentioned for your information.

Layout work involves the use of all of the measurement tools shown in Chapter 6. The materials used for marking the steel are also a part of this process. Study each print long enough to become familiar with the size and shape of the item to be fabricated. Read all notes. Look at the specification box (also called title block). Is there a revision block or box? Some prints may have a revision stamp on them. It is usually in red letters. This is good news because it means that mistakes have been corrected. It does not mean that even a pilot model has been finished. It just means that someone found errors that required the changes made on your copy of the print. Always check to see that subdimensions add up to the overall figure.

Since most steel requires the use of power equipment, find a place easily accessible to such machines. The longest shapes must be laid out first. Never cut a 40-foot (12.2-m) beam into 6-foot (approximately 2-m) lengths, only to find that several pieces of that material needs to be 34 feet (10.2 m) in length. That error may be worth your day's wages, and several such errors worth your job. Check each dimension twice and measure twice before cutting once.

When laying out hole centers, cross-check at punch-marked centers (Fig. 9-11). This would be centers A to D and B to C. Check again with the print to see that the layout on the 12-inch beam is correct. The depth of section on this beam should be 12.12 inches or 12⅛ inches. It would be to your advantage to obtain a small pocket handbook of steel weights and measures from your supplier. This is particularly true if you must pull this beam from a large stockpile. The local warehouse may have 20 or more sizes of 12-inch (30.075-cm.) W beams.

An excellent tool for this layout would be the combination square, particularly if an 18-inch (46-cm) blade is used. When set for 13 inches (approximately 32 cm), it will pick up points A and C. When set for 5 inches (12.75 cm), it will pick up B and D. By using a scribe and sliding the base of the square along the end of the beam, you can mark the steel with a readily discernible groove or scratch. If you do not need a permanent line, use soapstone or a silver marking pencil. To complete the other lines, set the blade for 10 inches (approximately 26 cm), and slide the base along the flange to produce line C-D. Repeat this procedure using the other edge of the flange to produce line A-B. The real advantage of soapstone over the pencil is that once it is

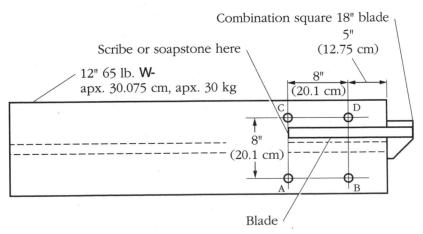

Scale 1½ (38.1 mm) = 1" (25.4 mm)

9-11. *Procedures for beam layout.*

marked and later dampened, the soapstone line will reappear when heated. Both the scribe lines and chalked lines are difficult to follow when cutting with an oxy-fuel torch. This is always true if the layout might be abraded before it can be put to use. A more permanent method of marking is to center-punch the lines at ½-inch (12.7-mm) intervals. This line is now very easy to follow, but time is lost if it is not necessary.

If a permanent center point is needed, the punch mark should be very deep. If possible, it should be ³⁄₃₂ inch (2.3812 mm) deep. It cannot be produced by a spring-loaded automatic center punch. If you circle this mark with a permanent type paint, it will be easy to find. A case in point would be when you need to lay out the same radius or arc on many pieces of steel. The center point of a tank bottom should have such a mark, especially if a soapstone point is used with trammels. The scraping of the side plates as they slide into position may wipe out base line marks. This problem is eliminated if butterfly guides are placed on each side of the base line with just enough space to accommodate the thickness of the first course of side plates (Fig. 9-12). The same type of guides are used to align and to land steel. This does present a problem for the fabricator, who now must accurately position the sections and tack weld them in place. The base line (molded line) is in the center and under the steel just put in

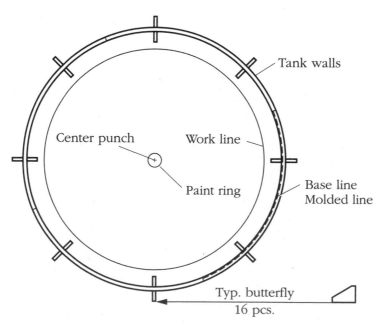

9-12. *The use of molded and work lines in tank and bulkhead fitting.*

place. This too is solved by the use of work lines, which are also shown in Fig. 9-11. The work line is usually center-punched. The automatic center punch has sufficient force to do this job. Make three punch marks approximately ½ inch (12.7 mm) apart and skip about 12 inches (approximately 30.075 cm) and repeat. This system is always used in steel boat (ship) building when bulkheads or complete sections are to be placed by cranes.

The framing square was first used on wooden structures, but it was instantly adopted by steel fabricators. Although steel may take many shapes, the square and rectangular sections are the ones most used. Angles, channels, and beams all lend themselves to these structural forms. Highrise buildings, crossmembers for large and small trailers, bridge support members, and many other steel fabricated items use these two geometrical shapes. This square fits neatly into the webs of channels. It is a natural 90-degree angle and easy to use with one hand, leaving the other hand free for marking. It lines up the flanges of channels and beams and the legs of angles. The figures etched in its blades show rise and run (pitch), for sloped structures and how to calculate the rise and tread dimensions for stairs. The information is also given for proper cutting lengths and angle degrees for cutting angle braces and brackets. It can also be used in conjunction with the level and plumb- bob to provide right-angle changes of direction and to position cross members in hard-to-reach areas. All layout people need this tool.

The combination square is used in two other distinct layout procedures. The center head attachment fits the same blade, and when the legs touch the edges of a circular piece of steel or a round shaft end, the blade automatically lies on a perfect center diameter of the piece. If you move the legs approximately one quarter of the way around the circle, the blade now crosses the exact center of the piece. This layout procedure is used in many ways. One use is to find the center of the king pin (the male part of the automatic locking device of the large trailer hitches; Fig. 9-13). Once the center of the pin is found, use a plumb-bob to drop a line to the floor. Now find a spot at the end of axle as close to a wheel as possible. Measure the distance to the plumb-bob point. Find the safe spot on the opposite end of the axle. If that distance is the same, lock the whole axle in place and recheck your measurements. This triangular method of alignment will ensure the perfect tracking of trailer on the highway. This is also the correct way to position a hitch/axle relationship on a small utility trailer for home use. If two or more axles are required, you must know the size tires to be used and the clearance needed between the tires. Once this is found, you can put an adjustable lock on one side, tying it to the frame.

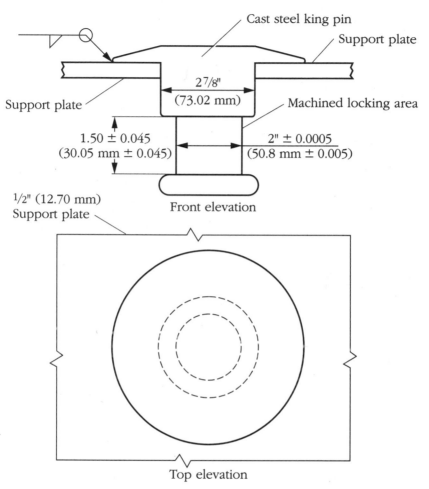

Cast steel king pin

Support plate

Support plate

27/8"
(73.02 mm)

Machined locking area

1.50 ± 0.045
(30.05 mm ± 0.045)

2" ± 0.0005
(50.8 mm ± 0.005)

1/2" (12.70 mm)
Support plate

Front elevation

Top elevation

9-13. *A type of hitch device for large trailers.*

Cut a length of steel that touches the outside circumference of both axles (the shortest point between them). The steel shape best suited for this is ½-inch (12.7-mm) round stock, but any-easy-to-handle ridged shape will do. The industry calls this a *story pole.* You can use it to set the other end of the axle. If you triangulate the points touched by the story pole on the axles, you will have a complete story and a trailer that tracks perfectly. You can also position the axle or axles in this manner with the frame of the trailer and then establish the spot for hitch or pin center. This simply reverses the whole procedure. This one use of a story pole may bring to mind many places where it would be a perfect measuring tool. It works well where any parallel sections are to be positioned.

The use of floor layout is detailed in Project 6 of this text, and some of the layout previously mentioned is also touched on in that project.

Patterns are used where you need to make many pieces all alike. They eliminate the need for repetitious layout. Patterns save both time and material. By moving one around on a sheet of steel and fitting it next to a like piece or even other parts of the same thickness, you may be able to use almost every square inch of material. Some pieces may actually use the same *kerf* (the width of a cut using any conventional fuel or cutting process). To some extent, computer programming use, as shown in Chapter 7 for multitorch equipment, is cutting into the use of patterns in large shops. Perhaps your company knows that the investment in such a machine is not justified. In that case a cost study should be made to see if your steel supplier can do the work. Another factor that may be present in such a decision is if parts made from patterns will be needed in the near future and fabricators with great skill would otherwise be laid off. Consider such decisions carefully.

The layout of a large, complex, four-sided structure may require that opposite sides made of plate match each other. Make only one layout of each of the sides facing each other. When the parts for each such side are aligned and tack welded, use *these* parts as patterns. In this way the sides may not be perfect, but they will match each other.

Jigs are designed to hold parts in positions so they can be at least tack-welded. In some cases they will hold the pieces during the entire welding operation, which cuts down on layout time and also tire warpage and distortion that might result from the unrestrained welding of such parts. They may also ensure that all finished products are similar and perhaps interchangeable. Be sure that slag and weld spatter do not impair the operation of the jigs or tolerance control.

Project 6: Making patterns

The sample of a useful pattern is the aim of this project. Figure 9-14 shows the pattern. It is not a jig; it holds nothing in position. This pattern uses any scrap of 8-inch (203.2 mm) standard black pipe. About a 2-inch (50.8 mm) length of pipe will do the job. Six pieces of light sheet metal (any thickness) will be satisfactory for use. They should be 1⅜ inches (34.925 mm) wide and 12 inches (304.8 mm) long. They should be welded to the outside of the pipe as shown in Fig. 9-13. This pattern is used to mark the areas to be removed from a pipe for a welder qualification test. The sheet metal is positioned so that one strip can be used for the top pull test coupon.

9-14. *A pattern for a pipe layout.* Lane Community College

The second strip is exactly opposite for the other pull test. These two can even be narrowed to ¾-inch (19.5-mm) sections. The other four strips are positioned to mark the root and face bend sections of the test. Remember, the root and face bend coupons may be rounded by grinding. No torch marks should show. The pull coupons should not be rounded, but no torch marks are allowed. Any serration will start a crack. The same type of pattern can be used for any pipe diameter.

You can probably devise a pattern that will slip over various thicknesses of plate to mark the welder qualification coupons. Watch for the differences where side bend coupons are used.

Questions for study

1. Name five questions you should consider before starting a fabrication process.
2. What is a floor layout?
3. In what position should a steel beam be placed prior to being cut with an oxy-fuel torch?
4. Why does steel need a primer coat of paint as soon as it is cleaned after fabrication?
5. Name 10 skills you must have if you want to open a one-person fabrication shop.
6. What are some of the jobs of management personnel?

7. What are the skills a draftsperson will use?
8. What happens when an estimator's bids are too high or low?
9. Why should the finished products be inspected?
10. What might a foreperson see that a fabricator can miss, and why?
11. Should you choose a partnership over a small corporation?
12. Of what advantage is an excise tax exemption?
13. Why would you want insurance for your company?
14. How important is safety to you?
15. What should you do if you notice unsafe conditions?
16. What is a rigger? What does one do?
17. Where can you obtain rules of safety for shop and field work in your state?
18. When should you discard used wire rope?
19. What types of equipment need safety guards?
20. When will you use some kind of safety equipment?
21. Name five safety requirements when using compressed air tools.
22. How many safety measures can you think of regarding the transporting, moving, storing, placing, treatment, and use of compressed gases?
23. Can you interchange hoses for different types of gases?
24. Should you tilt and roll cylinders on their bottom edge?
25. When would you lift cylinders by their protection caps?
26. What is meant by *cracking* a cylinder?
27. Why are electrode holders insulated?
28. What rules apply when welding or cutting around paints, fuels, or heavy dust concentrations?
29. When welding galvanized steel, what element is a problem? What can result from welding or cutting such material?
30. What toxic element is present in older painted surfaces?
31. When old fuel tanks are completely dry, what causes the explosion from cutting or welding processes?
32. What filler metals could contain cadmium?
33. When should you use respirators?
34. How many people can give directions to a crane or hoist operator?
35. What is the "head and shoulders" rule for safety?
36. Who should attend safety meetings? Why?
37. What should be the difference between sketches and blueprints?
38. What are blueprints?

39. How many views are usually shown on a blueprint?
40. What are the alternate names the views might have if an engineer draws the prints?
41. What views must be clearly marked if used on a print?
42. How should a center line be marked?
43. What does a break line show?
44. Why should you take a course in advanced blueprint reading or estimating?
45. What is involved in making a quick bid on a small job?
46. If a shop charges $54 an hour and you make only $15 an hour, how many other people make it possible for you to have a job, and how many other factors account for some of the other $39?
47. What is the weld symbol?
48. What does an arrow side symbol tell you?
49. Why was the flag symbol chosen over the large black circle?
50. Does a revision stamp on a print mean that the print is now perfect?
51. Why are the largest pieces or shapes noted on a print to be cut first?
52. Why would you indent a scribe line or chalked line with center punch marks?
53. Where would you use a combination square? A framing square?
54. When would dogs come in handy? How about butterflies and cranes?
55. What is a story pole?
56. When should you consider the use of jigs?

10

Fabrication practices

Steel-fabricated large structural items are all around us. You are familiar with many of these fabricated materials. You readily accept those you know. Skyscrapers all have massive steel support columns. These columns are generally stated in steel warehouse catalogs as W or W-F beams. The steel companies may also give a listing for H columns or bearing piles. Standard practice for such manufacturing companies is to supply beams or columns up to 50 inches (approximately 130 cm). This number refers to the depth of section. Farwest Steel's catalog lists W beams to 33 inches (approximately 90 cm) and gives design data for flange and web thickness, as well as flange width and weight per foot. Some of the beams show a cover plate that extends beyond the faces of the flanges.

The American Institute of Steel Construction has published some good books filled with answers to structural problems. While much of this data deals with making buildings, they also contain information on how the beams should be connected and detail the codes and specifications that have proven to be the best in the world. These manuals were produced for mechanical engineers, but much information on rivet and bolt strength and welded joint applications is of great advantage to you. If one of the fields of fabrication that you choose is steel erection, then the knowledge of bolt sizes, lengths, and strengths is absolutely essential. Most of the buildings now being erected do not have a single rivet in them, but many of the bridges do.

Steel bridges have been around for a long time. Nearly 100 years ago, welding was sent to the blacksmith shop (forge welding) and even oxy-acetylene welding was not the smart way to join steel. As previously stated, riveting of steel has some distinct advantages over welding. Perhaps you could relate to them better in terms of plug welds. There is no danger of crater cracking. There are no square corners where stress cracks can originate. They can give slightly without losing any strength. If you work in the field as opposed to a shop, you may need to have some bolt-up skills. The field jobs always pay better than shop jobs. If bridge repair is required, you may need to replace rivets as well as bolts.

Riveting is both a skill and an art. We have shown some of the skills you will need to master in previous chapters. Before we again note those and many more not previously mentioned, you should look at why the repair is needed. Did the rivets fail? Were too few rivets used for the beam-to-plate or beam-to-beam connections? Were the legs of angles of sufficient thickness and length to support more rivets? See Fig. 10-1. This sketch is not intended as an engineering blueprint. It is meant to show the material and weight that would be added for a riveted joint or seat bracket. It also shows the complex problems of rivet placement. Consider the need for a large compressed air gun and a bucking (backup) gun. It would require a pneumatic tube for a delivery of the rivets and a rivet forge or oven to heat the rivets prior to delivery. The guns require large-volume, compressed-air hoses. Consider the extra steel involved, the time spent in drilling and reaming the holes, the individual procedures for the job, and the three-person team needed.

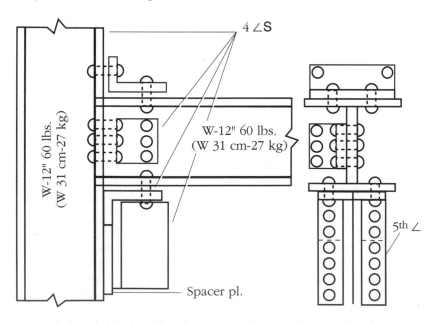

10-1 *A detail sketch of the placement of rivets when used in beam connections.*

The bolting of the same sections would use a similar amount of steel and drilling, but it could be done without the reaming process and, of course, the rivet heater and that person's equipment. There would still be a need for the skills of two persons, one to handle the bolt-up gun on the nuts and one to hold the bolt heads. It is plain to see why welding is the answer in almost every case. For the joint shown in Fig. 10-1, we could eliminate the angles and spacer plate. We could do without the drilling and reaming of the many holes. If a beam seat bracket was required, a 14-inch (38-cm) length of 8 - × - 8- × -½-inch (20.5-cm- × -20.5 cm- × -12.7-mm) angle might be the choice of the engineer. Of course, you would weld the vertical beads, tying the angle to the column and the angle to the edges of the beam flange only (Fig. 10-2). Never weld across the face of a flange except where it ends at that point and the engineer has so designated.

Figure 10-2 shows a typical knee brace and a gusset reinforcement, which may be acceptable alternates for welded connections. Again, the engineer will have to decide the load and stress factors involved. These calculations may take into consideration concrete facia as stiffening or concrete slab flooring 5 or 6 inches (127 or 154.4 mm) thick, as shown in Fig. 10-3. This reinforcement may be steel straight rods running the full length of the slab. It can also take the form of grids placed at different levels as the pouring of the concrete takes place.

Steel catalogs can you very good figures of sizes, grades, and number designations for those sizes. As a case in point, these reinforcing rods (hereinafter called *rebar*) will be given with their own values. A #3 will have approximately a $\frac{3}{8}$-inch (9.525-mm) diameter and a yield point strength of either 40,000 pounds (approximately 18 metric tons) or 60,000 pounds (27 metric tons). This figure is per square inch. The #18 rebar has a diameter of approximately $2\frac{1}{4}$ inches (57.15 mm). The reason these diameters are given as approximations is that the rebar is hot-formed and serrations are stamped into the surfaces of the rebar so that concrete can form into the irregularities for better holding power. The grids mentioned earlier can be welded or wire-tied when they cross each other to form various square and rectangular shapes as required. Welds should be of filler metals that match the grade properties of the rebar.

Reinforced concrete is formed into both columns and transverse members with slab coverings. The transverse connectors have the rods pulled into torsion while the concrete is poured and held until it sets. This type of construction was used in highway overpasses that broke up in the earthquakes that have plagued the California highway system. Now circular grids are used for the full height of the

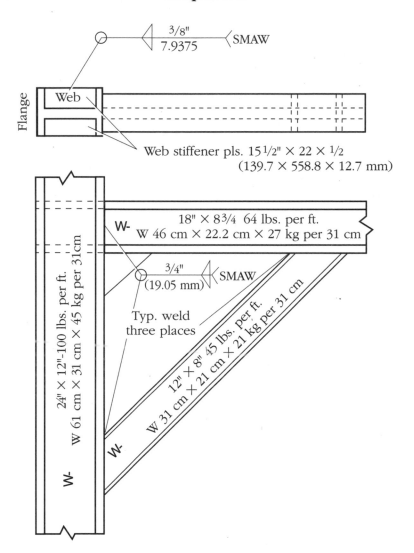

10-2 *Placement of welds for a beam connection.*

columns. New construction techniques have been engineered to pre-vent such problems. In these same quakes, all steel bridges showed some twisting and other minor problems, but, to my knowledge, none failed or caused loss of life.

The bridges in the United States are in sad shape. One study said that 38 percent of the bridges in the nation needed repair or replace-ment. This must be done in a timely manner, but it will cost hundreds of billions of dollars. While computer-aided engineering can solve the problems and estimate the costs, men and women will have to do the

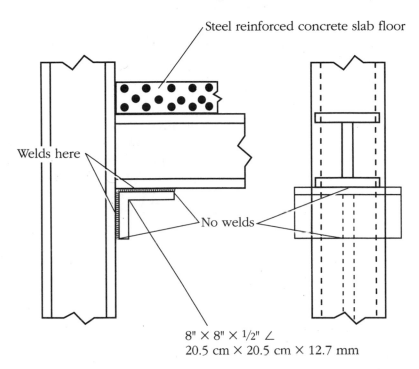

Steel reinforced concrete slab floor

Welds here

No welds

8" × 8" × ½" ∠
20.5 cm × 20.5 cm × 12.7 mm

10-3 *A cross section of a concrete slab with the placement of steel reinforcing rods.*

fabrication and erection work. The working conditions on bridges and highrise buildings are terrible. Someone calculated that at 200 feet in the air, it takes three-quarters of your strength to stand in place. At the 200-foot (approximately 81-meter) level, ground wind velocity doubles. Fear of heights is very real for most people. This fear is a healthy thing, because it makes you very careful.

All tools must be carried in a belt pack or put on tethers as mentioned in the safety section in Chapter 9. The small chain or rope must bear the weight of the tool and the drop impact load as it reaches the end of the tether. The safety margin for such material should be a personal matter. A ¾-inch (approximately 19.05-mm) spud wrench (the measure refers to the jaw opening to fit a nut for a bolt) weighs about 3 pounds (approximately 1.375 kilograms). If dropped 200 feet, it will go through a hard hat and kill a person instantly. Even a ¾-inch (19.05-mm) erection bolt, which weighs 0.986 pounds (approximately 2.3 kilograms), will do great damage if dropped 40 or 50 feet (12.2 or 15.2 meters).

You may need special clothing for working in rainy or snowy weather. You need boots, perhaps with steel toe caps, a cold-weather cap to fit under your hard hat, safety goggles, and perhaps ear plugs. If you are welding, get a hard hat/face shield combination or a hood that can attach directly to a hard hat.

While buildings are usually privately financed, union help is often the choice of contractors. A skilled labor pool is available in most areas. If additional help is needed, these unions can call in help from other cities or even other states. If you are temporarily unemployed, check with the union that uses people with your skills. Highway and bridge construction is always government-financed, and the unions are required by law to put your name on the out-of-work list and call you in the order in which your name appears. The problem here is that multiple skills may be needed to qualify for a particular job.

There are several good things about this type of work. For almost all welding jobs in field work, certification is required for each person. All welding personnel are offered testing at the contractor's expense. Usually the contractor pays an hourly rate for the time you spend in welding and preparing the test coupons (specimens). This process is detailed in Chapter 12 of this text. If you are successful, certification papers will be supplied to you that show you are a fine recommendation for subsequent employment. Of course, you also get the job, and the pay is almost double for that of shop work. If shop pay in your area is $11 an hour, then field pay will be about $22 per hour.

If you must work outside of the city where you are recruited (the contractor and/or union will set a prescribed distance), then you will be paid *subsistence*, a given amount for food and lodging. The company may choose to pay all such costs, as well as travel expense for the first trip to the job and the last trip back to your home base. Generally, this pay amounts to between 30 and 50 cents per mile. The company also pays a fixed amount into a retirement fund, about $1 for each hour worked. Sometimes a vacation amount is set, about 50 cents per hour. If you are required to work more than 8 hours in one day or 40 hours in a week, union contracts call for double-time pay. This is also true of any work on a holiday. You will soon realize that $44 per hour is a good wage regardless of working conditions. Accident insurance is required, and you may have to give a token amount (a few cents per hour), but the contractor (erector) pays all other costs.

After you work out of a union hiring situation for a short time, you may be asked or forced to become a member of that union. The initiation fee may be set at a few dollars or as high as several hundred dollars. A monthly dues charge is paid for union expenses. If you are not a union member yet, you may have a fixed amount taken from your check to pay the union for expenses and contract negotiations.

For no good reason, these fees are called *doby fees*. After you have worked out of *the hall* for about 3 months, the union may also supply health insurance for you and your family, which could include dental and vision insurance. Be sure you keep a daily account for all travel and subsistence pay. These items are not usually subject to state or federal taxes. Check with an accountant on whether safety clothing and equipment are also tax-exempt. If you go back and review that portion of Chapter 9 dealing with safety, you will find that most erection work requires a personal safety harness for all workers.

The beams and girders will be prefabbed. The holes will be drilled for quick attachment to the support columns. Some personnel will stab the sharp end of a spud wrench into a hole and guide the section into place. The crane operator and the head rigger must work as a team to inch the sections into place without touching anyone. A beam weighing 50 or 60 pounds (23 to 27.5 kilograms) per foot that is 40 feet (12.2 m) long has a life of its own when it swinging on the end of a 50-foot (15.2-m) wire rope cable. That ton will smash any scaffold and push anyone from a fixed position.

While positioning a beam for attachment to a column by this method, you may also, by leverage on the spud wrench, force alignment of bolt holes. The erection bolts (temporary) may be of several lengths depending on the thickness of the matching flanges. This length must be sufficient for a full thread through the locking nut. If washers are required, that still does not alter the full thread rule. If the flange on either or both beams is tapered (mill or S standard beams), *hill side* washers may be used to even the pull on the bolt when it is tightened. If an uneven torque is applied, the bolt body may be scored or gouged, weakening the strength of the bolt. An erection foreperson or supervisor may receive a bonus for the amount of steel *hung* on his or her shift. This factor must not affect the safety of those around you or yourself. Check for the correct number of erection bolts in any connection and the full thread rule before moving to another spot.

If tack-welding is the temporary approved method of joining, the length and weld size must be watched carefully. Never leave a partially finished weld on a scaffold hanger (support member) or on a connection while you and your partner take a coffee or lunch break or over a shift change. Most companies allow a time to complete safety work, put away tools, and reach ground level by the time the shift ends. If the workers are union personnel, one person is a designated job *steward*. Any violations of safety standards, personnel problems with supervisors, and contract provisions regarding wages, hours, work load, speed of work or other matters should be brought to the steward's attention. By striking the steel with three sharp, ringing blows, the steward can bring the work to a complete stop. It will not start again until all matters are resolved.

On long-span bridges ($\frac{1}{4}$ mile or 0.4 kilometers and longer), you may be eligible for a hazardous pay scale for dealing with certain materials and specific heights. Hazardous pay scales are also used on smokestack erections and dam construction. Bridges are divided into two general classifications: support or suspension and standard. The classic example of support-arc hanging bridge is the Golden Gate Bridge in San Francisco. The massive support cables are of woven wire rope encased in steel tube, for easy paint maintenance. As previously mentioned, no steel subject to salt spray and wind and water erosion will last long in such environments. Painting crews have a permanent job. They start at one end and when they reach the other end it is time to start again. When crews first started painting that bridge, the durable paint of that time was so toxic that the life expectancy of a painter was seven years.

The cables of all suspension bridges are placed over huge tower structures of steel that are seated down to bedrock. From these big cables, smaller cables or rods are dropped down to support the cross-member beams just below the depth of material in the highway. The small bridge shown in Fig. 10-4 and the larger bridge in Fig. 10-5 are standard tower support types with stronger beams running both ways between the towers.

10-4 *A small steel bridge.*

10-5 *The center section of this bridge can be raised by motors to allow ship travel.*

The towers in all cases may be a group of steel columns encased in concrete or driven to bedrock. The bridge in Fig. 10-5 is a type of drawbridge; the center section can be raised for ship passage. It actually is a section of arched construction spanning a fairly large river.

The penstock (large, formed steel pipe) that carries water from a reservoir to a hydroelectric turbine is usually a part of dam construction. The penstock is formed in half sections. The rolling or press-brake operations must be close-tolerance processes, which are detailed in Chapter 8. The short half sections are then welded together to make a complete pipe section. The use of low-alloy, high-strength steel is a good choice of material. The first ones used mild steel (low carbon) that formed easily and welded with ease. The filler metals were 6010, 7010, or 6011 category. These numbering designations are explained in Chapter 11. The steel thickness varied depending on the diameter of the pipe. It often ran from $\frac{3}{4}$ inch (19.05 mm) to $1\frac{1}{2}$ (38.1 mm). The penstock I.D. varied from 7 feet (2.175 m) to 22 feet (6.75 m). The roundness of it had to be guaranteed to $\frac{1}{8}$ inch (3.175 mm) or less.

If welded into pipe shape, the sections had to be supported by temporary stiffeners while being transported. The stiffing support sections were called *spiders*. They were simple cross-bracing of Ls supporting the pipe at eight points around the inside walls of the pipe. These spiders were spaced out at different intervals, depending on the diameter of the pipe and the wall thickness. The (tin-can action) flexing of pipe as it was transported over rough terrain tended to expand the side walls, forcing an out-of-round contour. Now the penstock pipe can be brought in by helicopter.

The use of a *spreader* bar, as shown in Fig. 10-6, would help. This device might also be used to spread the load in any lift where a cable or sling would put too much pressure on one part of a fabricated structure. From steel weight tables, a theoretical penstock section could be a 12-foot (3.7 m) O.D. × 3.14 (pi) × 10 feet (3.075 m) in length × 40 pounds (18.1 kilograms) per foot. If the wall thickness is 1 inch (25.4 mm), then it weighs 15,373.44 pounds (7.6867 tons or approximately 7 metric tons). The 8-inch (21-cm) W beam would support the load even if measured at a given point. Since the shackles could be removed from the spreader temporarily, and the section slid into place, the full length of penstock section would be supported. It could be locked in place with set screw clamps for transport. An alternate method would be to use two standard plate clamps attached to the shackles on the bottom of the spreader bar, which would divide the load weight while securing the section. Standard plate clamps have a 5-to-1 rated safety factor.

For penstock placement, the sections must be anchored in place. On exposed penstock, concrete base material must be tied to deep rock filler or solid rock. The base material is cradle-formed to fit the diameter of fabricated sections. All fabricated sections are referred to as *cans* until attached. If the penstock is placed in a dam structure, the upper-end placement is critical. The reservoir should have sufficient water to cover the intake tunnel at all times, providing a steady supply to the turbines located at the bottom end of the penstock. If several turbines are in place, then branching of the penstock will occur as it emerges from the face of the dam. To house such a penstock, a tunnel is left in the dam structure in as much solid (native) rock as possible. The cans are fitted with outside retainer rings at given spacings (Fig. 10-7). The tunnel walls are 10 to 12 inches (25.5 cm to 31 cm) larger than the O.D. of the penstock. The bottom of the tunnel is fitted with high-profile railroad rails. The cans slide down the tunnel on the rails and into position. They are welded only after alignment is checked to 0.013 straight-line tolerance by transit. The contractor's figures and the Army Corps of Engineers results must agree before any welding is started.

The spiders remain in position until welding is complete. The cans are beveled to provide a 60-degree included angle with a ⅛-inch (3.175-mm) root gap opening. A ⅜-inch (9.925-mm) backing strip is tack-welded in place to the O.D. of each joint. All welding is done

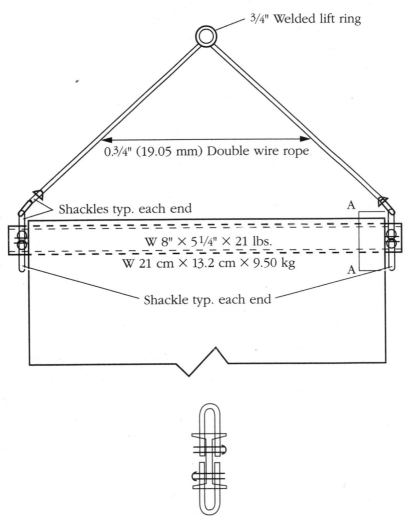

3/4" Welded lift ring

0.3/4" (19.05 mm) Double wire rope

Shackles typ. each end

A

W 8" × 5 1/4" × 21 lbs.

W 21 cm × 13.2 cm × 9.50 kg

A

Shackle typ. each end

Section A-A with **W** flange notched for shackels

10-6 *A spreader bar is used to support material that might be damaged as it is hoisted into place.*

from the inside except for the backing strips. The welding is usually done by four welders working on opposite quadrants to minimize distortion. The welding is extremely difficult. The incline angle of the tunnel may be 30 to 45 degrees. Almost all of the welding is out of position.

The weld metal has a tendency to *bag* when welded in the over-head position. For the flat or down-hand position and using the same criteria as above, you will use 2.4 pounds (1.28 kilograms) of

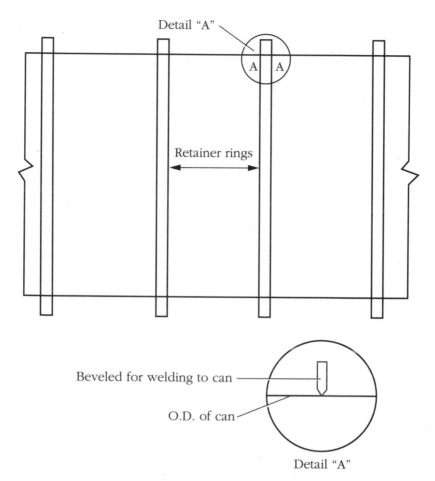

Detail "A"

Retainer rings

Beveled for welding to can

O.D. of can

Detail "A"

10-7 *Steel retainer rings attached to the outside walls of the penstock act much as rebar when concrete is poured around them.*

filler for each foot (31 cm). You need one-third more filler for this type of welding. Use two welding caps, one standard and one with the bill facing backward to keep weld spatter from hanging on the heavy weld jacket collar. Even heavy-duty welding mitts last only one week. Try cutting a large section from an old mitt or glove. Cut a hole just large enough to slide over the electrode holder or gun grip. This works well when welding with large-diameter cored wires too. When the welding is completed on a section, the welds are ground smooth. The force of the water going over an obstruction digs a hole just beyond it. This is the same thing that happens just behind a rock in a stream.

As a production welder, you will be issued a personal identity stamp. It will probably be a common steel letter stamp. If your stamp is the letter A, you will drive the A into the steel wall of the penstock at the exact spot where you start welding. When you reach a coworker's stamp, you put your stamp next to it. These stamp markings show up clearly on X-ray film. It is not necessary to do more than leave a light, clear letter. Inspectors frown on too deep an indentation. X-ray testing is required, and welding film is filed away after all repairs are complete. The contractor pays for the process and is responsible for all welding and materials. A welder is allowed one inclusion (porosity or particle) in each foot of weld. Five to six passes will be required for 1-inch-(25.4-mm-) thick steel. No inclusion shall be more than $\frac{1}{16}$ inch (1.5875 mm) in diameter or length. If slag inclusions or porosity exceeds this allowance, the weld passes must be removed, usually for 6 to 8 inches (15.2 to 20.075 cm) on either side of the offending material. No wagon tracks (incomplete tie-in with any weld passes or parent material) is ever allowed. When the repair is finished, it is again X-rayed. If the repair was successful, that shot is attached to the original for filing.

After all inspection is completed by the contractor and Army Corps of Engineers personnel, that phase is completed. Hydraulic cement then ties the penstock and retainer rings to the tunnel walls. The dam gates, used to release excess water or for penstock repair, are positioned as needed. They are fabricated in sections and brought to the dam site for connection and installation. Since millions of metric tons of material go into the dam itself, you may think of an endless chain of trucks bringing in rock. If the engineers choose the spot, they may position the dam where solid rock walls form one side of the dam. This rock may be a part of the dam itself. As mentioned earlier in this chapter, the tunnel for the penstock may be a part of the construction planning. If this is the case, some of the rock above the dam may be brought down by conveyor.

Conveyors are used in almost all industries where material or people need to be moved efficiently and quickly. You may not be required to fabricate moving walkways as used in most large airports. You will need to know about the fabrication and support of other conveyors. Figure 10-8 shows the type of conveyor used for bringing rock to a dam structure. Almost all mining operations use this type of conveyor.

Sawmills use conveyors to bring large logs to the barkers and haul the bark away. The rough green lumber is conveyed to sorting tables. The sawdust is sent to piles for the making of briquettes. The scrap lumber is sent to chippers when it is prepared for making chipboard or pressboard. Even postal companies use them to move packages. Food processors use them in every stage of operation.

10-8 *A belt-type conveyor for rock.* Delta Rock Products

Steel support members are always used where the materials conveyed are exposed to weather. As shown in Fig. 10-9, the conveyor supports are simple angle or channel frames. They hold the rolls, which are mass-produced. The rolls can follow any contour from flat to cradle form. They have a based structure for easy attachment to the support frames. The problems here are mainly relative to the weight of the material conveyed. The rolls are made with proven engineering guidance for any operation. The ones used in Fig. 10-10 are standard for rock-crushing operations. They have a load factor of 4 to 1. They certainly will wear but will not collapse.

Wet sand is probably the heaviest material handled in such operations. The loading may demand large support members. The longitudinal side sections could use large channel or even M beams. Some contractors have used box sections made of light angle crossbraced with the same material. The cases of failure were largely due to insufficient columns of tower support. The spacing on the towers may be 50 or more feet (approximately 15.2 m) apart. The top tower must be extra strong. It must stand the load plus the weight of the drive motor and gear reducer, the drive, and the drum. The pulley must be smooth to accommodate material to 3 inches (7.6 cm) average depth across the width of the belt. Dry gravel weighs 100 pounds (approximately 45.5 kilograms) per cubic foot. On a 4-foot-wide (approximately 1.23 meters) belt, each running foot will hold 100 pounds (45.5 kilograms) of gravel.

10-9 *Typical support members for conveyors.* Delta Rock Products

The base support for the tower is a concrete footing with a base plate of steel. The steel plate is sufficient size to stand the load and is placed in the form as the concrete is poured. It will have anchors of rebar welded to the plate as shown in Fig. 10-11. The larger-diameter rebar must be formed using a radiused die or hot-formed to avoid

10-10 *Rolls in sections support the belt and assist its travel.* Delta Rock Products

cracking. Several dams in the Northwest used huge belt conveyors to bring rock to the site. The endless belts were fiber and steel reinforced to withstand the gouging of large, sharp-edged rock. They were 6 to 8 feet wide (1.81 to 2.45 m). The towers were constructed as shown but often had four support columns.

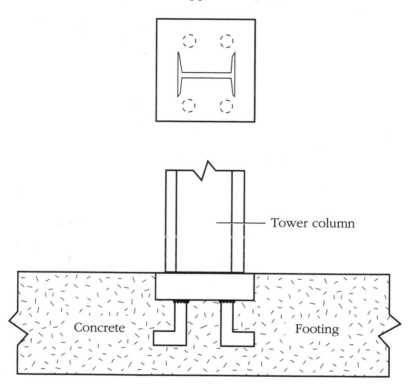

10-11 *The use of J bolts to anchor base plates.*

Storage tanks

Storage tanks come in all sizes and many shapes. There are completely round tanks that are pedestal-mounted or supported by columns at four or six points. The reason for such mounting is to provide gravity feed, rather than having to pump the liquid from the tank. The main problems here are the forming of plates, which must be rolled in two directions. The height of tanks from ground level produces many erection difficulties. The scaffolding to fit even the bottom curved plates must be tied to support columns. The welding and back gouging for second-side welding are difficult. Cranes are used to put up the steel and even to provide steady tie-off points for the safety harnesses of workers.

Most of the welding is out-of-position work. If the cylinder could be constructed at ground level and lifted into place as a unit, most of the erection cost would be eliminated. In any case, holding bands might need to encircle the tank. These bands can provide better support spread over a wider area. The band or bands could be welded to the tank using a chain or staggered intermittent fillet weld system. The bands would still have to be welded to the support columns. The testing, if done at ground level, would be much cheaper.

Never construct a square tank for transporting even small amounts of liquid. Some tanker trucks used for forest-fire protection in the Northwest overturned when this type of construction was attempted. Baffles (partitions) with round flow-through holes do stop some of the shock-wave action in the tankers you see on the highway. In the square tanks, the water and weight shifted to the side as the truck rounded a corner. Also, the force will try to spring the plates and tear out corner weld seams.

The making of storage tanks become a highly specialized fabrication and welding process. See Figs. 10-12 and 10-13 for flat-bottomed tanks used the world over.

The same type of plate pattern is used for tanks and ships. The checkerboard placement of plates may be varied, but the practice of offsetting each seam is strictly maintained and for the same reason. The weld quality may be two different code standards, but for the

10-12 *A storage tank with a flat bottom and an access or drainage opening.*

10-13 *A large circular storage tank with girth seams clearly visible.*

latter type of tank mentioned, the welding is almost always an approximation of API specifications 1104. The radiographic procedure of testing is used so that a permanent record can be kept for each weld.

No large self-supporting tank is ever of square, cubic design. The pressure of fluids or continuous side seams make such a shape impractical for the same reasons previously given.

All plates for circumferential use are preformed (rolled) to exact contour. The *cast* is maintained in transit to the erection site by cradle-cribbing of wood blocks or beams.

The flat bottom plates are held to a diameter approximately 6 inches wider than that of the tank proper. These plates are over-lapped at each seam by about 2 inches. The welds, then, are all edge welds, but the plates are only *production-tacked* until all other welds on the tank are made.

The girth plates are maintained on edge in vertical position, usually by the use of turnbuckles and cables. Because of problems in handling, most plates do not exceed 20 feet in length and 8 feet in width or height. Thickness is dependent only on the height and diameter of the tank. The columnar pressure on the side walls is evenly distributed and not of a magnitude you might expect. The first *course* of plates on a million-gallon tank usually does not exceed ½ inch in thickness. As the subsequent courses are added, plate thickness is reduced until a ¼-inch thickness is reached at the top. The toadstool is a center column usually standard pipe that rises to a height slightly higher than the rim of the outside circumferential walls of the tank. The ribs cap the pipe and radiate out to a ring of angle or channel formed to the contour of the tank and used to reinforce the top wall and support the top plates. The top plates are usually pie-slice-shaped segments that lap each other on the ribs (often called the spider) and also each other so that butt welding is not required.

Spacer bars are set at the end of each girth plate. No beveling is done on the *verts*. The spacing allows the welder to reach deep into the open square butt joint. Welding is done with a view toward weld integrity at least through half the joint thickness. The arc-air process is generally used to groove the reverse side of the joint. (When welding the first side, the joint opening is reduced by shrinkage and slag from the root of other side weld. It must be removed before the welding of the joint can be completed). Each vertical joint is always supported by strongbacks on one side of the joint. If the joint sags inward or outward, the strongbacks are used to correct the inclination and perhaps overcorrect if it is suspected that first-side welding will cause joint distortion that will not self-correct by weld completion. Wedges used in conjunction with the strongbacks and the ever-present saddles are the tools for the job.

The diameter of the tank is a *ring line* or *molded line*. See Fig. 9-12. The thickness of the first course plates is divided by that line. *Butterflys* are set far enough on either side of the molded line to allow for wedging the plates to conform to almost-perfect tank contour.

As soon as the second course verts are welded, the first circumferential welds must be completed. If extra courses are added before girth welding is done, plate sag and overlap will always occur, and if continued weight is added, the tank will self-destruct.

As the tank is completed, the bottom plates are finally production-welded (the tacks allowed for some lateral creep to minimize locked-in stress). The last welds are the bottom ring welds (fillets) inside and out. Testing of the lap welds used on the top and bottom plates is usually done by vacuum process and is deemed adequate for standard storage tanks. Testing of verts and girth welds are usually done by radiographs, but identifying stamps are not often used for individual welders.

The fabrication of trailers has become a very competitive field. Small utility trailers are often mass-produced. The entire framework, including cross members, are placed in a jig. The tongue and hitch assembly may be included in the same jig. All of the channel sections are precut by chop or cut-off saws using abrasive wheels rather than steel circular blades. With special attachments, the iron worker can clip off the channels as rapidly as they can be positioned. The abrasive circle saws can cut the miters for tongue parts. Once the members are tack-welded in position, they can be moved to fixtures or positioners. The production welding can be done by manual GMAW (gas metal arc welding). It could also be done by robotics.Lightweight recreational boat trailers are manufactured in this manner. They may use an angle-iron frame structure. It is difficult for a small shop to compete with such construction methods. The only factor that might offer advantage to a small shop would be shipping costs.

If your shop decides to go into specialty products, you should consider air reserve tanks. These tanks come in a limited number of sizes. All heavy equipment trailers, including semi-trailers used nationwide, must have air-brake systems. The tractor truck has a small air compressor unit run by the motor of the truck. The compressed air in the tank locks the brakes, causing the tires to drag rather than roll. A steel warehouse can supply stack-cut 14-gauge (approximately 0.07 inch or 1.8 mm) sheets in exact lengths and widths. Simply roll it to a finished diameter and then seam-weld it. The ends are *swedged* (necked down) by dies. Other blanks of like steel thickness are formed into domed heads to fit the swedged ends. Before the heads are welded, make sure that holes are stamped into the tank bodies and head blanks to eliminate vacuum pressure as the weld metal cools. The heads can then be position-welded. Finally, threaded pipe fittings are welded in place for attachment to the air lines. Because of shape, these items are not easily shipped in a cost-effective way.

The frames of heavy-duty, highway-use trailers are usually made of aluminum or low-alloy, high-tensile-strength steel. The aluminum is always alloyed for strength, which does add a little weight and cost. The low-alloy steels add a little cost but may cut the weight of the frame by as much as two-thirds. By using these steels, the welding costs are greatly reduced, compared to the welding of aluminum alloys. The common flat-bed trailers are not truly of flat design. When not under load, the arching of the main frame members is clearly evident. The amount of arch is a matter of engineering and load requirement. The methods of cambering (arching) steel were fully detailed in Chapter 8. The use of gussets as previously shown transfers some of the load to the cross members, strengthening the arching main line sections. The turntable, self-locking fifth wheel connections is universally used. It is easily positioned by the tractor operator because of wing approaches to guide the *king pin* into this connector (Fig. 10-14). Once the king pin strikes the back of the fifth-wheel structure, a spring-loaded pin snaps into place at the front of the machine-locking area. It holds the king pin so there is little or no movement. The support plate is welded to a box section composed of channel cross members that strengthen this area. A hole is cut in the center of the support plate to accept the 2⅞-inch (7.30-cm) body of the king pin. It is then welded to the support plate as shown correctly in Fig. 10-14. This is a fillet weld all around. Since the king pin is cast and the support plate may not be mild steel, choose a low-hydrogen filler metal.

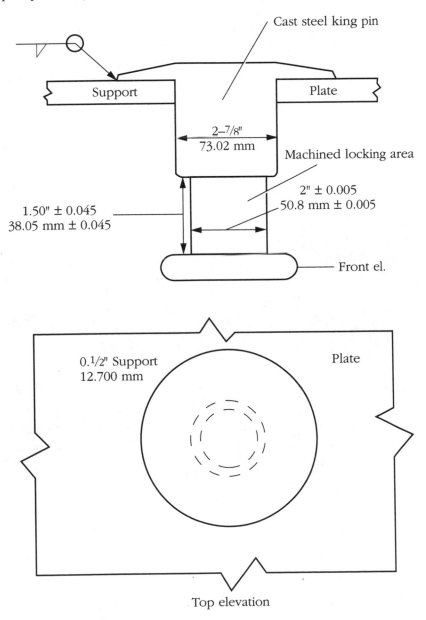

10-14 *The male half of a hitch for heavy-duty trailers.*

The *low-boy* (low-profile main body) of this trailer is always cambered to support and transport heavy equipment. The trailer shown in Fig.10-15 can haul a 40-ton (36.2-metric ton) load easily. If the load is over width (8 feet or 2.475 m is the absolute limit for legal highway use), then outriggers can be used. For this you will need a special highway permit. It may require flag vehicles at the front and back to advise motorists of the overwide load. The outriggers are hinged or socket- mounted for easy removal. They attach to the main frame and generally accept one 12-inch (30.2-cm) by 3-inch (7.6-cm) rough green timber or plank, allowing the trailers to haul motorized shovels and cranes with wide tread design. The trailer in Fig. 10-15 has a narrow-body main frame. It also has permanent outside frames to accept high-profile equipment. Such items may have either track or wheel design. High-strength, low-alloy steel is always used in the manufacture of such trailers. The uses of these steels were outlined in Chapters 3 and 8.

10-15 *Notice the cambering (arching) of the trailer support members.* General Trailer Co.

The General Trailer Company of Springfield, Oregon, manufactures heavy-equipment trailers used worldwide. The patented load boosters shown in Figs. 10-16 and 10-17 may commonly be called *jeeps* and *tagalongs* by the heavy-equipment industry. The frame booster uses the same gooseneck design as the main trailer in some cases. The back boosters are more of the more common flat contour.

In all such trailer devices, the need remains the same. By using up to 11 axles and 42 tires, the load weight is spread over a greater distance, making possible the legal payload weight of more than 78 tons (approximately 70 metric tons). This company also produces dockside container trailers. These are wide-body trailers with several main frame members. They may use up to 24 axles and 100 tires. The trailers remain on the docks and transport many massive containers to and from dock warehouses. They were fabricated in sections and assembled at the work site.

10-16 *A load booster, an add on trailer dolly that spreads load weight over many axles and tires.* General Trailer Co.

10-17 *The triple-axle spread with 12 tires and two-hitch arrangement make this a popular booster.* General Trailer Co.

The need to transport nuclear waste was solved by using honey-comb wall construction of the containers. The containers were suspended on gimbles on the trailers. In testing, the trailers were struck by locomotives at legal rail speed. The trailers were damaged as calculated, but the containers had no damage due to the superior design of trailer and containers.

Figure 10-18 shows the proper way to connect the gooseneck to the main frame. Never stop a frame plate at points A or B. Other companies tried this and the trailers usually failed at one of these places. The web plates should never be cut as shown by the dotted line.

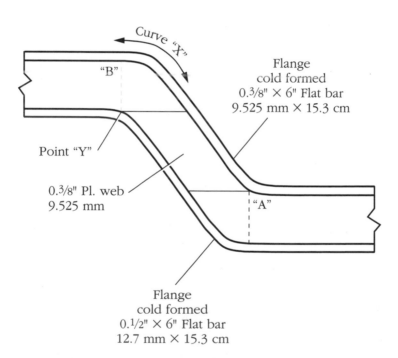

10-18 *The web plates of the fabricated beams are all parallel to the load and road surfaces.*

If a splice is necessary in the flanged flat bar, it should be made as detailed in the upper part of Fig. 10-18, never as depicted by the dotted line. The splice should be double-beveled for complete weld penetration. After the first side is welded, the root should be ground to solid metal. Then the second side should be welded. Care must be taken not to undercut the bar at the weld edges. After the excess metal is ground to a smooth flat surface, it is ready for fabrication. Calculate the amount of flat bar needed for the top flange. Do not splice it once part of it is already welded to the web. Repeat this procedure for the bottom flange bar. If you use thick bar for these flanges, it will be necessary to prebend the material in a press. On a gentle curve, exact bend position is not as important as when an abrupt change in direction is required. See curve X and point Y in Fig. 10-18.

Fabrication of ships

While I will make no attempt to cover all the fields of the shipbuilding industry, I will address the main fabrication techniques. Figure 10-19 shows an ocean-going vessel. The *ways* are skeletal structures that surround and support the *boat*. Remember, it remains a boat until it is commissioned.

10-19 *A Liberty ship built in 1943. Many of these ships are still in service today.*

The *keel* (bottom) plates are patterned for the type of ship needed. Most ships are not flat bottomed but contoured to reflect curvature in one or more directions. The *draft* or *draw* of a vessel is directly related to this curvature and refers to the amount of vessel below the water line and can be stated in terms of loaded or un-loaded status. The *bow* is the front of the vessel (Fig. 10-20) and is more curved than the *stern,* or back, of the vessel. The *starboard* side of a vessel is the right side as you face the bow while on deck. The *port* side is to the left.

10-20 *View of the bow of the ship being loaded at a dock.*

As the flatter plates give way to those curving up around the sides, they become *hull* plates. The hull is the main body of the boat. It encompasses everything from the keel to the *gunnels* (an archaic term for gun-rail), usually just a few inches above the top or main decks, except for two heavy plates or castings. These are the *society* plates, which tie the hull plates together at the bow of the boat, and the *wrapper* plate that serves the same function at the stern.

The fabrication and prefabrication processes start as soon as the ways are complete. The ways, usually a wood form, are an outside skeleton frame work. At the front is a massive concrete block. It has two or four large flat bars extending from it. When the first keel plates are laid (put in place), one or more bars are attached to those bars extending from the concrete block. They will hold the boat in place until it is launched. The keel plates will be double-beveled before being laid. The joints will be *faired* as shown in Chapter 9.

Production tacks, or intermittent welds, are used. The tacks are 3 inches (approximately 7.6 cm) in length. The *pitch* (distance between weld centers) is generally 15 inches (approximately 40 cm).

That means weld 3 inches and skip 12 inches. These tacks are about ³⁄₁₆ to ¹⁄₄ inch (4.76 to 6.35 mm) thick. Before final production is begun, all slag in removed from the tacks to leave only bright metal, both at the root and the face of the tack. A ¹⁄₈-inch (3.175-mm) root gap is maintained for most keel and hull plates.

Strongbacks, shown in Fig. 10-21, are generally placed at 2-foot (62-cm) intervals for the entire length of the joint. If there is space under the strongback, except at the rathole, a saddle is placed over the strongback and welded to the low spot in the steel. A wedge is driven on top of the strongback, forcing the low spot to rise to meet the strongback along its full length. The wedges may be tacked to the strongback to hold them in place until all welding is complete.

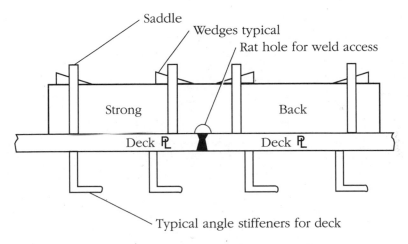

Saddle
Wedges typical
Rat hole for weld access
Strong
Back
Deck ℀
Deck ℀
Typical angle stiffeners for deck

10-21 *A strongback, saddle, and wedge configuration.*

The double bottom sections are brought in as prefabricated structures as soon as the first hull plates are production-welded to the keel plates. The rib angles may or may not be attached to these first hull plates. It depends on how the tank top deck plates are put into place. As soon as the tank top is secured to the bottom interstices and hull plates, the bulkheads are landed. The butterfly system is used. These sections are kept in place by the use of wire rope cables attached with clips. The clips are welded to the bulkhead and to the tank top. A turnbuckle is usually placed between the tank top clip and the cable to allow fore or aft movement. Since position of the bulkhead is critical, the shipfitter must use care. The between-deck sections are prefabricated. They are set in place with the hold openings ready-framed. They have escape hatch and safety ladder sections attached. These decks may also need to be fitted using many of the tools shown in Chapter 6.

The hull plates are joined as pictured in Fig. 10-22. The top deck may be welded or even riveted to the hull at the gunnel area, which could allow some movement to help compensate for wave action. Fuel storage tanks, fresh water tanks, and refrigerated food storage compartment are usually located on the second deck. The super-structure is also prefabricated and brought in by two large gantry-type cranes. The bridge structure is then added. The masts are added on the main deck, directly over a bulkhead, for support. The masts may have radar as well as radio equipment installed in or on them. Small storage lockers may be installed on the main deck for easy access by the crew.

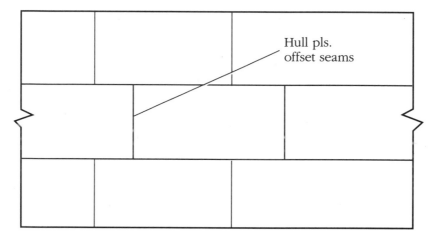

Hull pls. offset seams

10-22 *The offset or staggered plate formation eliminates inline weld seam cracks.*

The main hull is divided into sections called *holds*. The length of the vessel usually dictates the size and number of holds required. The holds are separated by *bulkheads*, just as walls separate rooms in a house. Just above the keel plates, a complex honeycomb section of small compartments make up the *double bottom*, and this stiff, re-silient section is capped by the *tank top*, which ties and seals to the hull. The tank top supports the bulkheads (extra stiffening plates in the double bottom are added at this point), and the bulkheads tie and seal to the tank top, the hull, and the top deck. Heavy angle iron shapes are used to stiffen the hull between the bulkheads. These may be called *ribs* and serve much the same function as the ribs of a per-son. They are welded to the hull in a continuous or intermittent se-quence and to the tank top and top deck. Brackets (usually triangular-shaped steel-plate gussets) are used to stiffen and transfer load from the ribs to the other parts they tie to. The ribs are usually not more than 3 feet apart. Often replacing a rib at 20- or 30-foot intervals is a large fabricated shape called a web frame, which is added for additional strength and support. The angle iron shape is also used to stiffen bulkheads.

The hatch, a small square or rectangular opening in the main deck, is used for loading cargo into the holds. It is framed and stiffened by angles under the decks, leaving the top surface smooth. Figure 10-23 shows a typical cross-sectional area of a vessel.

Plate thickness may vary greatly, but cargo vessels often have hull plates ranging from ¾ to 1 inch (19.05 to 25.40 mm). The plates are always double-beveled all around so that if proper sequences are followed, welding stresses will be relatively equalized. Seams between plates are never continued beyond one plate width (see Fig. 10-22). The offset seam joining stops any cracks or stress continuities. Plates are *faired* (leveled with each other) using dogs and wedges, comealongs, and jacks shown in Chapter 6. All seams are strongbacked (temporarily stiffened to prevented warping and distortion during the welding process).

Since the ways are slanted so the vessel will slide smoothly down into the water, you would have a problem when using a level. That can be eliminated by using a declivity board. Since *targets* (zeroing points for transit aiming) are permanently affixed to the ways, it is simple to check level and declivity of major bulkheads and other parts.

Project 7: A scale-model round tank

From Table 10-1 showing U.S. gallons in round tanks, fabricate and weld a scale model of a tank capable of storing 100,000 gallons or more of liquid.

Example: Using a scale of 1 inch equals 1 foot, find the value for a tank 29 feet, 3 inches in diameter. Since each foot of depth would hold 5026 gallons, 20 feet of depth would hold 100,520 gallons.

If 14-gauge mild steel sheet metal is used throughout, it will simplify the fabrication and welding processes. Since we want to approximate the tank-building processes, shear the sheet steel into 8-inch-wide strips. Four strips 31 inches long will produce a base or tank bottom. Scribe a radius on this plate after finding the center by snapping a chalk line corner to corner. The centering process works on any square, rectangular, or trapezoid shape. It might be advisable to tack-weld the strips prior to this process. You will probably choose a microwire process. Use wire at least 0.02 inch in diameter, but not more than 0.035 inch.

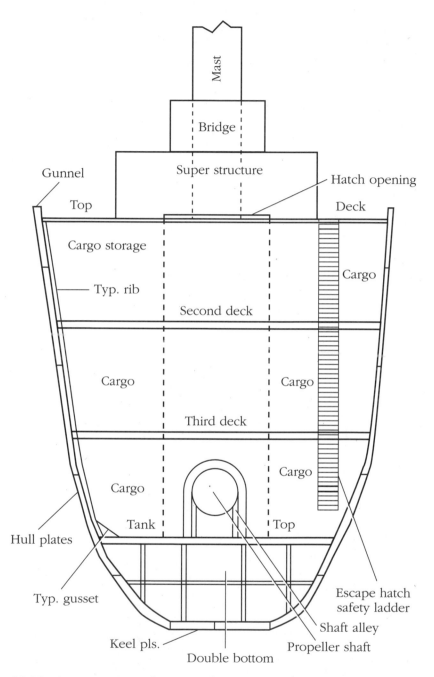

10-23 *A cross section of a cargo ship.*

Table 10-1. U.S. gallons for round tanks

 STEEL SERVICE COMPANY, INC.   **T-19**

Eugene   Roseburg   Medford

U. S. GALLONS IN ROUND TANKS
FOR ONE FOOT IN DEPTH

Dia. Tanks	No. U. S. Gals.	Cu. Ft. and Area in Sq. Ft.	Dia. Tanks	No. U.S. Gals.	Cu. Ft. and Area in Sq. Ft.	Dia. Tanks	No. U. S. Gals.	Cu. Ft and Area in Sq. Ft.
Ft. In.			Ft. In.			Ft. In.		
1	5.87	.785	5 8	188.66	25.22	19	2120.90	283.53
1 1	6.89	.922	5 9	194.25	25.97	19 2	2177.10	291.04
1 2	8.00	1.069	5 10	199.92	26.73	19 6	2234.00	298.65
1 3	9.18	1.227	5 11	205.67	27.49	19 9	2291.70	306.35
1 4	10.44	1.396	6	211.51	28.27	20	2350.10	314.16
1 5	11.79	1.576	6 3	229.50	30.68	20 3	2409.20	322.06
1 6	13.22	1.767	6 6	248.23	33.18	20 6	2469.10	330.06
1 7	14.73	1.969	6 9	267.69	35.78	20 9	2529.60	338.16
1 8	16.32	2.182	7	287.88	38.48	21	2591.00	346.36
1 9	17.99	2.405	7 3	308.81	41.28	21 3	2653.00	354.66
1 10	19.75	2.640	7 6	330.48	44.18	21 6	2715.80	363.05
1 11	21.58	2.885	7 9	352.88	47.77	21 9	2779.30	371.54
2	23.50	3.142	8	376.01	50.27	22	2843.60	380.13
2 1	25.50	3.409	8 3	399.88	53.46	22 3	2908.60	388.82
2 2	27.58	3.687	8 6	424.48	56.75	22 6	2974.30	397.61
2 3	29.74	3.976	8 9	449.82	60.13	22 9	2040.80	406.49
2 4	31.99	4.276	9	475.89	63.62	23	3108.00	415.48
2 5	34.31	4.587	9 3	502.70	67.20	23 3	3175.90	424.56
2 6	36.72	4.909	9 6	530.24	70.88	23 6	3244.60	433.74
2 7	39.21	5.241	9 9	558.51	74.66	23 9	3314.00	443.01
2 8	41.78	5.585	10	587.52	78.54	24	3384.10	452.39
2 9	44.43	5.940	10 3	617.26	82.52	24 3	3455.00	461.86
2 10	47.16	6.305	10 6	640.74	86.59	24 6	3526.60	471.44
2 11	49.98	6.681	10 9	678.95	90.76	24 9	3598.90	481.11
3	52.88	7.069	11	710.90	95.03	25	3672.00	490.87
3 1	55.86	7.467	11 3	743.58	99.40	25 3	3745.80	500.74
3 2	58.92	7.876	11 6	776.99	103.87	25 6	3820.30	510.71
3 3	62.06	8.296	11 9	811.14	108.43	25 9	3895.60	520.77
3 4	65.28	8.727	12	846.03	113.10	26	3971.60	530.93
3 5	68.58	9.168	12 3	881.65	117.86	26 3	4048.40	541.19
3 6	71.97	9.621	12 6	918.00	122.72	26 6	4125.90	551.55
3 7	75.44	10.085	12 9	955.09	127.68	26 9	4204.10	562.00
3 8	78.99	10.559	13	992.91	132.72	27	4283.00	572.66
3 9	82.62	11.045	13 3	1031.50	137.89	27 3	4362.70	583.21
3 10	86.33	11.541	13 6	1070.80	143.14	27 6	4443.10	593.96
3 11	90.13	12.048	13 9	1110.80	148.49	27 9	4525.30	604.81
4	94.00	12.566	14	1151.50	153.94	28	4606.20	615.75
4 1	97.96	13.095	14 3	1193.00	159.48	28 3	4688.80	626.80
4 '2	102.00	13.635	14 6	1235.30	165.13	28 6	4772.10	637.94
4 3	106.12	14.186	14 9	1278.20	170.87	28 9	4856.20	649.18
4 4	110.32	14.748	15	1321.90	176.71	29	4941.00	660.52
4 5	114.61	15.321	15 3	1366.40	182.65	29 3	5026.60	671.96
4 6	118.97	15.90	15 6	1411.50	188.69	29 6	5112.90	683.49
4 7	123.42	16.50	15 9	1457.40	194.83	29 9	5199.90	695.13
4 8	127.95	17.10	16	1504.10	201.06	30	5287.70	706.86
4 9	132.56	17.72	16 3	1551.40	207.39	30 3	5376.20	718.69
4 10	137.25	18.35	16 6	1599.50	213.82	30 6	5465.40	730.62
4 11	142.02	18.99	16 9	1648.50	220.35	30 9	5555.40	742.64
5	146.88	19.63	17	1697.90	226.98	31	5646.10	754.77
5 1	151.82	20.29	17 3	1748.20	233.71	31 3	5773.50	766.99
5 2	156.83	20.97	17 6	1799.30	240.53	31 6	5829.70	779.31
5 3	161.93	21.65	17 9	1851.10	247.45	31 9	5922.60	791.33
5 4	167.12	22.34	18	1903.60	254.47	32	6016.20	804.25
5 5	172.38	23.04	18 3	1956.80	261.59	32 3	6110.60	816.86
5 6	177.72	23.76	18 6	2010.80	268.80	32 6	6205.70	829.58
5 7	183.15	24.48	18 9	2065.50	276.12	32 9	6301.50	842.39

31½ GALLONS EQUAL 1 BARREL

To find the capacity of tanks greater than the largest given in the table, look in the table for a Tank of one-half of the given size and multiply its capacity by 4, or one of one-third its size and multiply its capacity by 9, etc.

A gallon of water (U. S. Standard) weighs 8⅓ lbs. and contains 231 cubic inches.

A cubic foot of water contains 7½ gallons, 1728 cubic inches, and weighs 62½ lbs.

To find the pressure in pounds per square inch of a column of water multiply the height of the column in feet x .434

Although the sheet metal will be easy to handle and shear to the proper circumference, the plates for a real tank would probably be flame cut. On the circular base, scribe the actual tank size. Knowing that plates rolled would be about 20 feet in length, shear 10 pieces 20 inches and 3 pieces 12 inches long (92 inches per course). A pattern of cardboard or sheet metal should be made to pick up the curvature from your layout.

Run a piece of scrap metal through your rolls. Check for proper curve with the pattern. The first pass through the rolls will take maximum curvature. If more passes are needed to match the pattern, you will need to back off roll pressure slightly before running the other strips or plates. Check the curved piece to the base plate layout.

Tack the first course to the base plate. Tack-weld the second course on the first course. Make sure that each vertical seam or joint is offset at least 3 inches. Never line up weld seams.

On a real tank, you do not add a third course before welding all of the vertical (verts) first course seams. I have always run the first horizontal *girth* joints completely around the tank before adding any additional plates to the tank.

If the added weight or striking the tank as more plates are positioned causes lower tack-welds to snap, the plates would move out of alignment. Realignment is difficult and may even bulge the plates, causing rejection by inspectors.

Add all additional courses to the tank, welding as you go. The tank top may be flat, but in most cases, it tapers upward toward the center. This form is often referred to as the *Coolie hat* form. It may be stated as *pitch* or *rise and run*. The base plates and tank tops are usually the last to be welded.

You be the inspector. If you are satisfied with the finished tank, fill it with water. If there are leaks, you must drain the tank before you do any welding repair.

Project 8: Utility trailer

A comprehensive parts list is the correct way to start the fabrication of this trailer. The complete material list is in Table 10-2. If you check with local suppliers for cost and availability of the items needed, it will save time. You are now the purchasing agent. In every case, call two or more suppliers. Place all the steel in one column. Use another column for all material such as lights, wiring, reflectors, trailer hitch, jack, conduit, and related requirements. One of the steel supply houses will probably have the ready-made fenders as well. If there are no trailer supply warehouses near you, try the hardware and auto supply section of a large department store. This is a good place to purchase low-cost tires. The store will have information on the load capacity of the tires. Check any wrecking yard (used auto supply depot) for the wheels. It may also have the axles, hubs, bearing, and spring assemblies at greatly reduced prices. The more material you can purchase from one outlet, the better the price may be. The wood deck material will have to come from a lumber warehouse. If you can have all materials on hand before the start of metal cutting and other prefabrication work, it will save time. The lack of one item may make completion of the project impossible.

Now you become the estimator. Since the parts are small and require some fitting and installing processes, be careful. If this is a first-time process, remember to double the cost of materials (this amount now includes labor and miscellaneous dollars. You may wish to add an additional percentage for shop cleanup, painting, and other factors before setting a price).

The large fabrication shops that maintain their own parts department add 80 to 100 percent on all parts. Their overhead costs are large, and a small shop can compete very well. When all materials are available, lay out a chalk outline of your trailer. as described in Chapter 9. You may want to show all cross members in your layout. Cross-check your layout to make sure it forms an exact rectangle. You may also want to add the layout for the tongue to get the correct length and see how it passes through the front frame member (Fig. 10-24). You could make the tongue beam in two pieces and not pass it through the front member, but this is not usually done. It presents problems in aligning the two parts *exactly*. There are seven cross members shown inside the main frame. Remember the overall width of trailer cannot exceed 7 feet, 10 inches (240 centimeters), as measured to the outside of each fender. The stated length is 12 feet (380 cm). It can change to suit your needs. The width *cannot.*

Cut the channels, beams, and angles to the dimensions of your layout. The outside frame members are placed on the chalk outline. Check the rectangle by cross-measuring one more time. You might also use the framing square on each corner; then tack-weld the frame securely. You may want to turn the channel with flanges in to present a nice, finished appearance that does not trap mud or other debris. The cross members 2 through 7 can now be placed and tack-welded. The spring hanger brackets can be welded to the frame or frame gussets. Use the parallel method for setting these hangers as described in Chapter 9. The axles should now be attached. They should be at right angles to the frame.

Table 10-2.
Material and cost list for parts and labor on a 12-foot (3.75-meter) utility trailer

# Pcs	Qty	Material Description	Length per item	Total length	Lbs. per foot	Total weight	Cost per item
A	3	6" (157.4 MM) 6.7 channel	20'0" 6.1 m	60'0" 18.3 m	6.7 3 kg	402 lbs. 180 kg	$160.20 US
B	2	3" (76.2 MM) 4.1 channel	20'0" 6.1 m	40'0" 12.2 m	4.1 1.86 kg	164 lbs. 74.4 kg	$65.60
C	1	4" (101.6 MM) S beams	20'0" 6.1 m	20'0" 6.1 m	7.7 3.5 kg	154 lbs. 70 kg	$61.60
D	3	3" (76.2 MM) S beams	20'0" 6.1 m	60'0" 18.3 m	5.7 2.6 kg	342 lbs. 156 kg	$39.80
E	3	3" × 3" (76.2 × 76.2 MM) angle	20'0" 6.1 m	60'0" 18.3 m	6.1 2.8 kg	122 lbs. 68 kg	$48.80
F		Dual axle set with brakes, hubs bearing, seals, shackles, and springs	92" (2.4 meters)				$427.00
G	4	Wheels for 13" tires	Used		$12.50		$50.00
H	4	Tires – 185-80 R13	New		$24.99		$99.96

I	1	6 way receptical (lights)			$8.00
J	2	Tail lights			$22.50
K	4	Reflectors			$12.80
L	40'0"	(12.2 MM) Wiring for lights			$15.20
M	2	Safety chain-rated 2-ton (1.8 M tons)	4'0' (1.22 M)		$12.80
N	1	Trailer hitch			$15.00
O	8	2" × 12" × 120" finished lumber	Treated (weather resistant)		$100.20
P	1	Conduit for wiring	30'0"	(9.2 m)	$12.50
Q	100	Nuts and washers for 3" (76.2 MM)	30'0"		$56.60
		Carriage bolts	5/16" × 3"	(7.9395 × 76.2 mm)	
R	1	Trailer jack			$22.50
S	1	Set of fenders for dual axles			$50.00
T	1	Ball to fit hitch for tow vehicle			$9.00
U	8	Gussets and light brackets			$12.00
V	1	10% Safety cost factor for miscellaneous items			$129.00
				Total	$1,420.96

10-24 *The fabricated main frame of a utility trailer.*

Use a story pole to check the distance between them.The tires must clear the frame by 1½ to 2 inches (38.1 to 50.4 mm).

Once the axles are aligned, you can place the tongue member and pierce the front frame to allow for passage of the tongue back to tie in with cross member 2. Chapter 9 tells you exactly how to position the hitch socket. This part of the hitch is made to weld to the tongue beam. The stiffener angles or beams of your choosing tie the front main frame channel to the tongue when the alignment with the front axle is perfect (Fig. 10-25).

The gussets at each tie-in point strengthen the tongue against both side play and any twist that would occur under load. The first cross member is in two pieces that further strengthen the tongue and tie it to the main side channels. The fenders are attached to the frame with production tacks or intermittent weld procedures. Note the gussets at the front and back of the fenders.

The fenders should be positioned so that there is about 3½ inches (88.9 mm) of clearance between the top of the tire and the inside surface of the fender to allow for complete spring action. On rough road, the tire should never contact the fender. After you make the triangle check again from the hitch to the front axle, you are ready for production welding. When the welding is completed, clean away all weld splatter and inspect the frame assembly for missed welds. Install the tail lights, wiring, and protective conduit for the wiring. Be sure there is enough wire and conduit to attach to the tow vehicle. The reflectors attach to the sides of the main frame a short distance from front and back.

10-25 *The tongue alignment, placement, and gusset reinforcement of a utility trailer.*

In some states and other countries, you may need a small amber clearance light in place of reflectors, which requires more wiring. Any large trailer or auto supply company can supply complete *harness* packages to meet your needs. The trailer jack can now be installed if required.

The trailer is now ready for prime coat painting. Review the safety painting rules. Wipe down the trailer to remove any oil, grease, or other material. Spray equipment or *rattle cans* will do the job. Allow drying time and complete the painting with the finish coat. If the safety requirements in your shop did not mention protecting the surrounding area from *overspray*, do so anyway. Always wear a protective mask and clothing. *Never* spray in windy conditions. The overspray can carry many yards. You can install the wood decking as soon as the paint dries. Drill both the wood and the cross members at the same time. It may be necessary to attach tabs to the front and rear main channels to provide drilling points. The carriage bolts will sink into the wood when tightened, and only a rounded surface will show above the wood. If drilling tabs were necessary, you might want to repaint those places before the deck is installed. Attach the safety chains to the tongue assembly. Inspect the completed trailer and clean up the area. Remove all leftover paint to meet environmental standards. Check the labor cost (your hourly rate) against your estimate. (See Table 10-2). If the cost was somewhere near your estimate, you have done a good job on this project.

Questions for study

1. What is a bearing pile?
2. Who needs to know about bolt strength?
3. Where was most of the welding done 100 years ago?
4. What does HAZ mean to you?
5. Name three advantages that riveting of steel might have over welding.
6. When shop jobs are safer than field jobs, why would anyone choose a field job?
7. Why is welding the first choice for joining steel today?
8. What is a tether? Where is it used?
9. Where is a knee brace used?
10. What is rebar?
11. Can rebar be welded?
12. Why are steel bridges still a good choice for highway construction?
13. Do computers play a part in the construction and repair of bridges?
14. Name three personal items you will need if you work high steel?
15. Why are certification papers valuable to a person?
16. What are the reasons for joining a union?
17. What is the full thread rule?
18. What type of bridges are shown in Figs. 10-4 and 10-5?
19. What is a penstock?
20. Where is a spider used on penstock?
21. What would you need to make a spreader bar?
22. What is a retainer ring as used on a penstock?
23. The testing of welds on a penstock is done by what method? Why?
24. Name five places where conveyors are used.
25. Why are flat-bottom storage tanks easier to work on than a pedestal-mounted tank?
26. Why not use square storage tanks instead of round ones?
27. When are some plates welded using a lap system rather than a butt joint process on storage tanks with flat bottoms?
28. What are butterflys used for?
29. What types of material are used in the fabrication of heavy-duty highway trailers?
30. If the main body of a flat-bed highway trailer is not flat, what shape is it?

31. Can a lowboy trailer carry a greater load than a flat-bed trailer?
32. How much weight can a lowboy trailer with boosters legally carry?
33. When you splice a flange for a lowboy trailer, how do you prepare the flat bar and make the welds?
34. What is a bulkhead on a ship?
35. How are the angle iron stiffeners attached to the bulkheads and hull plates of a boat?
36. What tools are used to fair plates on a boat being fabricated?
37. How are the keel plates prepared before they are positioned on the ways?
38. Where is a pitch distance measured on an intermittent weld?
39. What are strongbacks and how are they used?
40. What is the makeup of a double bottom and where is it located?

11

Welding practices and procedures

The history of welding depends entirely upon the type of welding involved. A kind of welding was used to produce bronze religious statues in the Far East, including India, China, and Malaysia. Parts of the statues were cast and then assembled for welding.

The hammering process was used and, of course, the amount of heat used is unknown. It is possible that a kiln or oven was built around the entire statue. The parts then could have been separated by thin layers of tin bronze or even lead bronze. This could have been added after the parts reached a brazing temperature. The sheets of these lower melting point materials would have been dropped into place from slits at the top of the kiln. The slits would have been covered until the correct temperature was reached, perhaps just above 1000 degrees F (540 to 560 degrees C). It would not have needed 1200 degrees F (approximately 650 degrees C).

This method would be a bonding action, not a fusion-welding process. If you wanted to try a bonding process, you could do so in the following manner: Obtain some fluxed brazing rod from a welding supply house. Heat a small strip or plate of steel enough so that it reaches a purple color, almost a red hue. The oxy-fuel torch is fine for this purpose. The flux (coat on the rod) acts as a cleaning agent when the proper heat is attained; direct heat on both the rod and the plate. The crystal structure of the plate has now expanded. The spaces between the crystals have enlarged. The brazing material will follow the path cleaned by the flux. With a little practice, a nice bead will form on the top of the steel. The brass or bronze will enter only the top few air spaces between the crystals. As the cooling process takes place, the material will shrink, locking the filler material into the top of the steel. This is also called a cladding process (Fig. 11-1). Note that it is not a fusion weld.

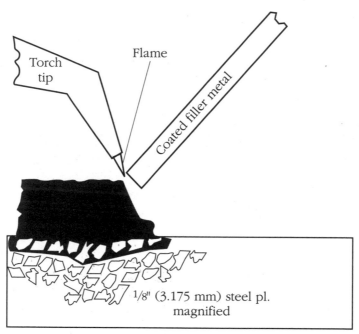

Torch tip

Flame

Coated filler metal

1/8" (3.175 mm) steel pl.
magnified

11-1. *The bonding or cladding process.*

The ancient craftspeople must have known a great deal about metal. The statues were formed more than 2000 years ago. Examples of the hammer welding of iron are carbon-dated back to around 1000 B.C. The tombs in Egypt have yielded weapons, some of which fashioned using this process.

An English gentleman, Sir Humphrey Davies, is given credit for finding the potential of acetylene when mixed with oxygen. Calcium carbide mixed with water produces acetylene. He also is said to have produced an arc between two carbon electrodes. One source says the *leyden jar* was the first battery. Sir Davies did use a battery to produce the arc. While some experiments were done by people in all parts of the world in the time period between 1850 and 1870, not much of note was accomplished until the last part of the nineteenth century.

Stanislaus Olszewaski, a Russian working in a French laboratory, was granted a patent for inventing arc welding in 1885. He also should have received a patent for the first electrode holder. It was carbon arc, and he did weld iron with this device.

Some welding of steel was done prior to World War I, but, as usual, it took a war to push the process forward. Prior work on filler-metal electrodes was done both in Europe and the United States. These rods were not coated at first. Some welding supply houses may be able to obtain a few bare wire rods for you. You will find that you cannot weld in corners with them. The arc will pull for the nearest ground. With a present-day motor generator welding machine, bare rod will do a good job on wire mesh screen. The slag cover produced by coated electrodes will do a bad job in similar cases.

The first U.S. patent for filler-metal electrodes was granted to C.L. Coffin in the late 1880s or the early 1890s. Many coatings of a lime base were tried to give the arc more focus. One such early product was water glass, which was really a potassium silicate. When mixed with water it becomes somewhat adhesive. The rods were dipped in it and simply allowed to dry. The coating of rods became a bad joke. A rod was covered by an asbestos tube or sock. It was then wound with red ribbon to hold the tube in place. This type was sold commercially for several years. This rod was around long enough to have two common names: *barber pole* and *peppermint stick.* Some welding supply companies may keep a sample around for display purposes only. I'm sure you can imagine what would happen when asbestos was burned in the arc action.

Before the American Welding Society adopted a numbering system, each manufacturer had a way of identifying their electrodes. One company would use a blue dot on the butt end of a rod and three brown dots half way up the coated surface. Another company would use two pink stripes on the butt end and one green dot on the flux coat. A third would use a purple dot on the end, a green stripe on the butt, and a black stripe on the coat. If the rods were removed from their containers or mixed with others, you needed to be a brand inspector.

A clue to the present numbering system probably came from the Murex Corporation, which developed a fine electrode in England. It had the stripes and dots, but it also had a number. That number was 8016Q. From those numbers you can now tell how to use it successfully. The Murex Corporation was purchased by the Lincoln Electric Company some time after World War II. You can deduce from this number that the American Welding Society (hereafter referred to as AWS) decided on their system using numbers. Such a system fits in well with SAE and ASTM methods of classifying steels. The AWS not only developed an excellent and comprehensive guide, they worked with other countries to establish it worldwide. The only holdouts are some communist countries, including China.

It was also during WWII that the need for welding aluminum and its alloys came about. Prior to this time, all aluminum was joined by riveting. The system for numbering was adopted as these alloys were developed. While pure aluminum (#1100) is soft, it has a low melting point. In relation to steel, it also does not have great tensile strength. Now, the alloys 7xxx and 8xxx have tensile strength and armor qualities greater than many steels. Although this text is mainly for the fabrication and welding of structural steels, a few processes and fillers will be given for aluminum.

The war needs also caused the rapid development and use of stainless-steel filler metals. The welders in WWII were told when in doubt, use stainless. The 308 stainless was the steel available, and 308 electrodes were used for welding stainless and any steel that wasn't designated as mild steel (low carbon), including armor plate on ships and on armored vehicles. As previously mentioned, it was and is used for welding repair of Hadfield manganese. The 308 stainless filler metals have excellent elongation factors from 38 to 50 percent, which means they can expand under stress without rupture. So can the plain, low-carbon, manganese filler metals. The stainlesses cost three times as much. They cannot be cut with a torch, while manganese can. The manganese also floats impurities better, hardens under impact, and provides excellent surface-wear qualities while remaining ductile underneath. The early railroad rails had a manganese steel content. Perhaps that is why welding was first done with stainless electrodes.

The low-carbon steel electrodes of the WWII period were for use with one or the other type of welding machines. The ac buzz box had no excitor or rectifier. If the arc distance was right, and the sine wave agreed, you could start welding. If either one was off, the rod stuck to the steel or would not arc at all. The motor generators were far better during this period (Fig. 11-2).

11-2. *A motor generator, a welding machine similar to those used in the 1940s.* Hobart Bros.

There were exceptions to some good qualities. A couple of the major manufacturers of electrical machines had problems. If you run across any old motor generator welders at a giveaway price, check them carefully before buying. If a rated 300-amp welder does well on ¼-inch (6.35-mm) 6024 (iron powder) electrodes for a half hour, you have a good machine. The problem machines can barely handle ⁵⁄₃₂-inch (3.9687-mm) rods at the end of that time period. Hobart, Wilson, and Lincoln produced some very good machines at that time. If you recognize the names, then you know that their products are still some of the best.

The major producers of welding rods could not keep up with the demands of the war industries. Some rods came out *green*. The coatings were not completely dried. These rods stuck together in bunches. If you tried to separate them, some had excess coat on one side and others had a bare side. It is still possible that dampness in transit will produce a similar problem now. The supply house will be happy to know of the problem and take care of it. Don't change suppliers without checking. The electrode manufacturers will want to correct the problem and, in some cases, will supply additional material free of charge through your local warehouses.

Submerged arc process

Later in this chapter, the problem and correction of *fingernailing* will be covered in depth. Although the process of welding with continuous wire was invented in 1920 by P.O. Noble of the General Electric company, it was not widely used. In 1930, a man received a patent for the submerged arc process. He sold the patent rights to the Linde Company. It was sold under the trade name Union Melt. Some people thought that the Union Carbide Company had invented the process. This process is still one of the most cost-effective methods of welding steel.

From 1940 to 1945 the shipbuilding industry in the United States advanced the use of this process at a great rate. The shipyards in Portland, Oregon, used Columbia River (salt-free) sand as a cover agent. By now you have guessed that it was not underwater welding. The arc was started and the sand immediately covered the arc, protecting it from the atmosphere. The arc continued to fuse the steel plates using wires up to ⅜ inch (4.7625 mm) in diameter. The amperage ratings for submerged arc are very high. In one shipyard, they ran from 800 to 1000 amps. These were the machine settings, not at the arc. Since amperage dictates the burnoff rate, the deposition rate was 32 to 38 pounds (14.5 kg to 17.1 kg) per hour. With increased wire diameter and amp ratings to 1200, deposition rates above 42 pounds (19 kg) per hour have been achieved. These welds were for deck seams only, and the overhead welding was done first from below these decks. It is not possible to weld an open butt joint with this process unless backing is provided. The fluidity of the metal and the voltage required would force the weld puddle through the joint. Please remember that the voltage is force and amperage is the amount of current flowing in the circuit.

Transformer-rectifier machines are produced in sizes from 200 amps to 1500 amps. It is possible that you might want to consider connecting two lower-rated machines in parallel. This setup can be used in the arc-air process for metal removal. When cutting tabs on large castings or the castings themselves, you can use copper-coated rods to 1 inch (25.4 mm) in diameter or $^3/_{16}$- × -$^5/_8$-inch (4.7625- × -15.875-mm) flat electrodes. The 1-inch (25.4-mm) electrodes may require 2000 amps of current using smaller-diameter electrodes. This is probably the cheapest and fastest method of metal removal. Old weld removal and the preparation of cracked sections in thick material is accomplished with ease. The specially designed carbon holder and electrodes can be purchased from every welding supply house. The correct compressed air supply is from 60 to 100 pounds (4.2 to 7.0 kg per sq. cm). Once the arc heats the metal, the force of the air blows it out of the way. This is a fine way to remove stainless steel and hard-surfacing materials that can not be cut with an oxy-fuel torch.

Gas metal arc welding

H.B. Cary should receive credit for designing a good gas metal arc gun in about 1950. Up until this time, the expensive gases were always used. In 1953, both Novoshilov and Lyubavskii came up with the introduction of the carbon dioxide as a shielding gas. As constant-voltage welding machines came into use and smaller diameter wires could be used, the GMAW system was a success.

Cary's gun used a spool for the continuous wire. The spool guns of today use the same principle, with a nozzle to direct the gas coverage and a contact tube to energize the wire. The welding of aluminum was first done with such equipment because of the short distance from the spool to the arc. This method eliminated the kinking of the soft wire. The spool gun is a good way to weld cast iron and aluminum.

The use of flux cores in microwire diameters is still being improved upon (Fig. 11-3). The heat input and speed of travel are plus factors for this joining process. The use of the short-circuit dip transfer and short arc systems seem to all have come at once. The use of the larger wires and larger coils of wire increased the deposition rates and a large share of the filler metal market. Most outdoor construction projects use the GMAW process.

If wind conditions are present, it is easy to build a temporary wind break to preserve the gas coverage and the integrity of the

11-3. *Coiled filler metal wire (0.035-inch diameter).*

weld. Do not weld in the rain. H_2O means just what it states—two gases, both detrimental to molten metal. The hydrogen produces an unbelievable amount of porosity in the weld deposit. The oxygen attacks the molten metal, producing instant rusting action.

If we place two pieces of steel cut from the same plate outside atmosphere overnight, the action will be obvious. Weld three beads on one plate and leave the other plate as cut. The welds and the areas out to the edge of the heat-affected zone will show rust. The other plate will show none. The hydrogen porosity will cause many problems. If it is inline porosity, it will always crack when subjected to stress. The welds will be rejected if tests are done. This subject is detailed in Chapter 12. The use of flux-cored wire has greatly increased the deposition rates of the process, as evidenced by it receiving its own designation.

FCAW is a fine complement to GMAW. This wire is produced in following manner. A thin layer of steel just wide enough to produce the diameter of the finished tube is laid flat. This layer is called skelp. The powdered core material is added as tiny rolls produce the tube. It must be tight, but it cannot be unduly crimped as it is joined. The skelp may contain alloys to satisfy your welding requirements. The flux core may also add more alloys or just produce the shielding needed for the process.

Some of FCAW wires now produced can weld through their own slag, which is a real advantage when welding thick plate. Plate from ⅝ to 1½ inches (15.875 to 38.1 mm) in thickness can be welded at one time. Using a track setup and automatic travel speed, you can try this system. With a contact tube on each side of the horizontal arm of the motorized travel mechanism, you can run two wires. With a third clamped to the rear of motor housing, you can increase the deposition. Check with your welding supply house to find out which wires are best for welding through their own slag. The weld deposits are advertised as X-ray quality. That statement depends on the code that is being used. This topic is covered in Chapter 12. When using the submerged arc process, two or more wires can be used in the same puddle, and the flux covers all.

Although cored wire is more expensive per pound, it pays on many weldments. The standard gas metal arc welding process (GMAW) is generally used in microwire sizes. It was first called MIG welding (metallic inert gas welding). Since CO_2 is not truly an inert gas, the name was changed. GMAW can be used very effectively in making out-of-position welds. The $\frac{1}{16}$- and $\frac{3}{32}$-inch (1.5875- and 2.3812-mm) wires are used but are not popular with some welders. It does eliminate spatter, which cuts down on cleanup time prior to painting. The deposition rates are good, although some data may show 30 pounds (13.6 kilograms per hour). The usual rates are much below this figure. The cross-sectioned areas of various wires are the governing factors. The 0.020 wires for thin sheet steel using pulsed arc and a constant voltage machine may have a 1- to 2-pound (0.452 to 0.91 kilograms) rate of deposit per hour.

The $\frac{3}{32}$-inch (2.3812-mm) wire in the example may produce a true deposit rate of 18.2 to 22.3 pounds with a constant volt power source and 100-percent duty cycle. This is done in the down-hand position (flat). You may need a constant feed for the wire and full automatic equipment to achieve any such deposition rates. If you need to use larger wires, the common 500-amp welders will still be adequate power sources.

The larger guns may need water cooling. The CO_2 does provide some cooling action. If semiautomatic welding is used, the operators need clothing that protects against the heat and the rays from the arc. Never wear white or very light-colored, lightweight material. The same type of radiation projected by the sun is noted in arc welding. The ultraviolet rays will penetrate white loose-weave materials and produce very harmful burns. Such burns are known cancer-causing agents. Most welding helmets are curved to fit the face and direct the smoke from the welding process away from the nose and mouth.

The standard welding lenses come in various protective shades (Table 11-1). The amp rating at the arc and the arc length should be considered when choosing a shade for safe eye protection. If shades above 12 are not available, improvise. Using a flip front helmet, replace the clear glass next to the eyes with a cutting shade lens. Shade 3 and 12 make 15; shade 5 and 10 make 15. OSHA may not sanction it, but if you are welding with a 500-amp open arc, it may save you hours of misery.

The burn from arc exposure is very similar to snow blindness. The arc flash burn is not the split-second impact on the eye. The accidental arc strike or ignition may make you see spots, but if you correct the situation instantly, there should be no lasting damage. Always check your hood before starting to weld. The lens holders should be examined, particularly at the corners. The dark lenses must show no damage. The gold and silver lenses should have no scratches on the coat. Even a small nick on the corner of a lens lets in a lot of light. Lengthy exposure is the real problem. Small blisters form on the eyeball. The eyelid rakes across the blisters. The pain would be the same as having hot sand in your eyes.

If it happens to you or someone you know, recommend seeing a doctor immediately. Follow his or her instructions faithfully. If the eye is contaminated by dirt or infection-causing agents, it may be necessary to scrape the eyeball. The fabricator is more likely to get burns than the welder. If a welder is tacking sections for you or production welding next to you, shield your eyes in every way possible. Always think about safety. Review the work that you will do this day.

Table 11-1. Standard protective welding lenses

Darkened shade
Standard protective welding lenses

Oxy-fuel cutting light to heavy 3 to 6

Oxy-acty welding ⅛" to ½" (3 mm to 12.7 mm) 5 or 6

Oxy-acty welding above ½" (12.7 mm) 6 to 8

S.M.A.W. electrode sizes under ⁵⁄₃₂" (3.9687) shade 10

S.M.A.W. electrodes ⁵⁄₃₂" to ¼" (3.9687 to 6.35 mm) shade 12

S.M.A.W. electrodes above ¼" (6.35 mm) shade 14

G.M.A.W. steels light to heavy shades 11 to 12

G.T.A.W. steels light to heavy shades 11 to 12

F.C.A.W. steels light to heavy shades 10 to 15

Carbon arc air metal removal shades 12 to 15

As amperage ratings increase the darker shades of welding lenses provide eye protection.

The use of flux-cored wire has completely changed the hard surfacing industry. Most of the rock production equipment, including mining and earth-moving equipment, needs a coat of long-wearing material. Many wires have been produced to protect against impact as well as abrasion. Large earth-moving shovels usually have parts surfaced before any work is started. The bucket teeth, the lips (front edges of the bucket), the track pads, and latch plates are all cases in point. The blades for large earth-moving and clearing tractors are surfaced before use.

Some rock-crushing cone mantels (capes or outside coverings) are surfaced prior to use. The same is true of crusher rolls. These types of equipment require periodic resurfacing and repair. Many of the wear surfaces are of manganese steel. The FCAW process is well-suited for this type of resurfacing process. Since hand-surfacing materials should not be applied more than $\frac{3}{16}$ inch (4.7625 mm) thick, the use of build-up passes are required. In the case of roll crusher mantels, the shells are 1 to 2 inches (25.4 to 50.8 mm) in thickness. The wear should be even across the face of each roll, but it will never be because the rock tends to gravitate towards center (Fig. 11-4).

The companies that produced hard-surfacing materials all developed cored wire for this purpose. Victor and Stoody in the United States and Harfac in Canada were leaders in the field. The Canadian company was bought outright by the McKay company of Pennsylvania. A subsidiary of McKay Company was the Automatic Welding Company of Pennsylvania. The three companies developed fully

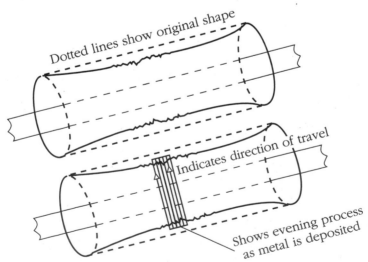

11-4. *The uneven wear pattern and the repair and build-up of rock-crushing roll shells.*

automated systems for the surfacing processes almost simultaneously. McKay used arc length to govern wire speed (deposition rate), which meant that as the wire left the contact tube, it was energized. It waited until the arc was established to send a message to the control unit. The distance of stick-out from the tube told the control to speed up or slow down the wire feed rolls. If a low spot, perhaps 1½ inches (37.5 mm) from the tube told the control unit send more filler, the rolls sped up. An advantage here was that wire was red-hot before ignition, requiring less amperage at the arc site. Because of this action, the low spots received more buildup than the high ones. The low areas might require two or more passes of manganese to reach a level almost equal to the edges. Since the roll ends have little or no wear, it is necessary to cover the hard overlay with a new layer of manganese.

Using a travel speed of 31 inches (approximately 80 cm) per minute, and full automation, deposition rates are excellent. A bead ⅝ to ¾ inch (15.875 to 19.05 mm) wide is achieved. The weld deposit is ³⁄₁₆ inch (4.7625 mm) in depth. Using a machine setting of 460 to 490 amps, and 36 to 40 volts, manganese can be run at 30 to 35 pounds per hour. The cored wire that produced these results was McKay 218-0. It was ⁷⁄₆₄-inch (2.7781-mm) wire. The tub or barrel was mounted on a turntable, which helped the wire rise smoothly from the container.

The wire could have been run from horizontally or vertically mounted coils. The advantage of using large auto-packs is apparent. These tubs and barrels contain from 100 to 500 pounds (45 to 260 kilograms) of wire. There is much less wire loss (the wire from the drive rolls to the contact tip cannot be used). Also the time lost in changing spooled coils of approximately 50 pounds (2.25 kilograms) is a factor. The hard overlay will be run at reduced rates and bead widths. The wire size may remain the same. Manganese-quality wires are also available in ⅛- and ⁵⁄₃₂-inch (3.175- and 3.9687-mm) diameters. Since increases in wire diameter increase the amp and volt requirements, you may not wish to purchase larger welding equipment. A 4-0 welding cable will carry 500 amps on constant potential machines.

All hard wire in the 30 to 67 Rockwell C scale hardness groupings stress-relieve themselves. Small cracks open at the weld surface. The amount of penetration is not great. While it does fuse with the manganese, the dilution of the hard material is minimal. If hard wire is added to more hard-surface material, the new follows a similar stress-crack pattern as the old. The two layers pull large areas free from the surface of the roll shell. This is called spalling. It should not occur if you follow the rule of never adding hard-filler metal to some previously surfaced steels.

If you often deal with steels of unknown hardness factor, carry your hardness tester with you (Project 2). A small glass or plastic tube and a ball bearing that moves freely in the tube is all that you'll need. Place the 6-inch (15.25-cm) tube in an upright position on the steel, and let the ball bearing fall down the tube. Mark the point at the top of the rebound. If you start with plain low-carbon steel and test increasingly harder steel, marking the rebound height with a scratch on the tube, you will soon have a good idea of comparative hardness of any steel. The ball will bounce higher as hardness factors increase. This will not tell you the type of alloys involved, but it will tell you not to weld with mild steel or very low-alloy filler metals.

There are many kinds of wear. The abrasion and low-impact wear mentioned in the examples are the most common. If water flow is involved, then corrosion and erosion are present, especially if particles are present in the water. Stainless steels may be the best answer, especially if drinking water is involved. If high impact is the problem, then very brittle materials should not be used.

Metal-to-metal wear should be easy to avoid. If one metal surface must contact another, and loading, moving, or both are present, dissimilar metals should be used. The friction of moving one plate on another under load almost causes bonding of like steels. Without using lubrication, galling (tearing action) takes place. We patented a method of locking phosphor-bronze plates between turntable parts on heavy equipment. This bronze has lubricating qualities. The wear that does occur is on the softer bronze plate and it is easily replaced if the need does arise.

The use of gas tungsten arc welding (GTAW) has been around a long time. The work of H.M. Hobart and P.K. Davies should be acknowledged. They used shield gases of argon and helium. A man by the name of Meredith patented the heli-arc process in 1941. Of course, tungsten does not melt when cooled by gas, and the arc is produced along the sides of this electrode in the welding of steel. Care must be taken in the sharpening of the tungsten. It should taper to a point. The length of the taper should be 2½ times the diameter of the electrode.

If the tungsten can be held to the stone against the diameter of travel and turned to perfect the taper, you will have a fine point. If you attempt to roll it horizontally to the stone, you may leave microscopic ridges on the taper that will cause the arc to flare outward instead of following a perfect line to the point. In most cases, GTAW processes do not use helium as a shielding gas. The helium lifts away from the puddle. That, of course, is an advantage in overhead welding. Helium also produces a hotter arc. The use of argon produces a more stable, better arc cover. If your equipment allows you to extinguish the arc and continue the gas flow over the weld puddle, it will protect the crater end. This process requires great skill and almost perfect welding conditions. The filler metal must be added and the puddle moved ahead as with the oxy-acetylene process.

Work in the shop is more suited to the process than in the field. Comfortable positions can be arranged. Wind should not be a factor. The transportation of gas is minimal. There is less wear and tear on the equipment. Finally, safety is easy to maintain.

If there is any leakage in the gas supply system, the atmosphere around us will be sucked into the lines and into the welds, which will cause instant porosity. GTAW is supposed to be an almost-perfect weld system. Most welds will be X-ray tested, and no porosity will be allowed. Because of cost factors and lack of skilled operators, not much low-carbon steel is welded in shops using this process.

The conventional and nuclear boiler industries have relied heavily on GTAW. The tubes that carry heated water and steam are of alloy steels. The root passes are generally made using GTAW. On conventional boilers, the subsequent passes are made with common low-hydrogen or chrome-moly (molybdenum) stick electrodes. The root pass welds are made using direct current straight polarity (DCSP). Two percent thoriated tungsten is the choice for most welding of this type. It holds shape better under most conditions. The tungsten electrode must not touch the metal being welded. If it does touch, it will cause a small amount of tungsten to be included in the weld, resulting in a hard spot (embrittlement). The tungsten will lose shape and have to be resharpened. It will not affect the rest of the filler metal. You should understand that in any welding process, even this one, all of the filler metal does not reach the puddle. Some of the material burns in the heat of the arc. The GTAW process and the inert-gas shielding provides the least amount of loss in the four major welding systems. The size of the ceramic nozzle is regulated by the gas coverage needed at the puddle. For tube welding on boilers, a short head set at 75 to 90 degrees from the handle may be used. The tungsten electrode may be 2 inches (100.8 mm) in length. With this hand piece, it is much easier to reach through the small opening between the tubes. It is often necessary to have a partner feed filler metal into the puddle, because you cannot reach the point needed (Fig. 11-5).

If stick electrodes are used for subsequent passes in such a situation, the partner will make a hot start, which is almost like lighting one match from another. As you reach a spot halfway around the tube, your partner positions his or her electrode to move into your puddle. With a little practice, this procedure is as smooth as passing a baton in a relay race. If the space is too small for the filler metal to pass between the tubes, it is wedged open above and below the joint.

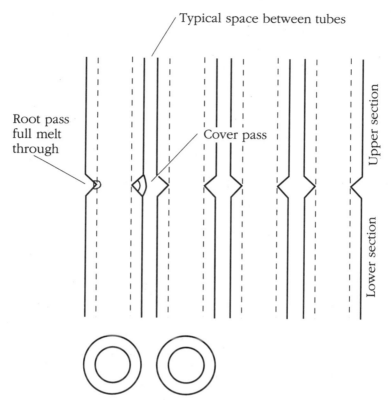

Typical space between tubes

Root pass
full melt
through

Cover pass

Upper section

Lower section

11-5. *Alignment and positioning of boiler tubes.*

If too much space is present between the upper and lower sections of a tube, this can be corrected. Heat the upper section slightly with an oxy-fuel torch while your partner pulls down on the tube. There is usually enough play to close the gap. If space still exists, repeat the process on the bottom section. Welders call this process smoking a section or single tube. The persons responsible for the shortage always hear about it. Figure 11-6 shows an older GTAW machine. At this time, new machines are being produced to cover all the variables that are now becoming apparent with this process.

While the processes covered above are part of the welding and steel fabrication practices, welding with stick electrodes is still sovereign. The shielded metal arc welding process should receive some credit for the winning of WWII. The U.S. industries were not damaged. They outproduced the output of the rest of the world by a three to one margin.

The aircraft companies did not use the process. Even the present-day electrodes are not used in the production welding of aluminum. The melting point is low in relation to that of steel, from 1200 to 1240 degrees F (approximately 590 to 605 centigrade). Even repair on thicknesses under ⅛ inch (3.175 mm) is not recommended. The heavy flux coat does not adequately protect the puddle.

In any noninert gas process, oxygen attacks the filler and the puddle. (In the steel-making processes, bags of aluminum are thrown into the furnace to reduce the harmful gases by attracting them and to promote their burning. Remember, the arc produced by the average electrode reaches a temperature of approximately 10,000 degrees F (5538 degrees C).

The mining of the bauxite (aluminum arc) was done by steel-fabricated equipment. This was true of the transported material. It was also true of the assembly process. Some planes were flown to staging areas during WWII; others were transported by ships. The welding of the vessels was 90-percent stick electrode production. The aircraft carriers utilized the same percentage figures. The troop transport ships and the supplies carried by the cargo vessels were key factors in the victory. Incidentally, the first cargo ships were called Liberty ships, but second groups were called Victory ships. Some cracking was evident in the ridged welded construction of the Liberty vessels. The notch effect on square corners of hatches and superstructure were the problem areas. The Victory ships were a vast improvement. The submarines and the surface war fleet were primarily stick-welded.

The locomotive and other railroad rolling stock supplied to Russia were fabricated here and shipped on the Liberties and Victories. The rolling stock was narrow gauge. Our rails are farther apart for better stability. Some of these old ships are still in service.

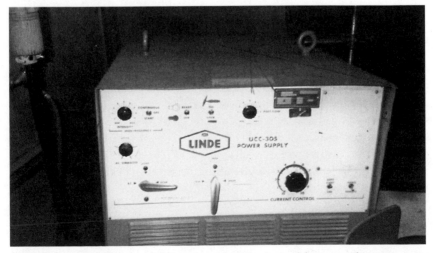

11-6. *A typical older-model gas tungsten arc welding machine.* Lane Community College

They sail under foreign flags and papers. Our codes and licensing procedures are much tougher than those of other countries.

Many of the early flux-covered electrodes are still being produced. The E-6010 and E-6012 rods are good examples. The E-6011 was developed for the ac (alternating current) welding machines. Transformers step up current; they do not produce power for welding. They are more quiet and do not use a constant level of electrical power. The fan is all that is running when the arc is not present. The ac welders do not have a serious arc blow problem. They are now produced with an exciter that steps up the sine wave flow to start the arc. With a selinium rectifier, they can go ac or dc (direct current). The E-6010s are not generally available in sizes beyond $\frac{3}{16}$ inch (4.7625 mm). The E-6012 can still be ordered in sizes from $\frac{1}{16}$ inch (1.5875 mm) to $\frac{5}{16}$ inch (7.9375 mm). It may still be possible to order 6010 and 6012 in even larger sizes. See Table 11-2 for general data. The E-6014 may no longer be produced. It is now superseded by E-7014. The E-7010 is available. The Shield Arc Company (a subsidiary of Lincoln Electric Company) developed this electrode. The trade name is Shield Arc 85. This is the rod listed with double numbers in Table 11-2. It was not brought out in 1985, nor does it have a minimum tensile strength of 85,000 psi (5950 kilograms per square cm). It is, however, the first choice of many pipe welders. With this electrode, there is no chance that even the highest values for grade B type F pipe are reached. The elongation figures mean added values in adverse weather conditions.

Let's look at the other extreme for the welding of mild steel plate. Some of the best welding engineers have stated that an E-4510 electrode would be the perfect match for A-36 steel. It is not listed in Table 11-2 because most filler metal manufacturers want an edge. All electrode producers cannot risk joint failure (where the steel does not fail, but the weld deposit does).

This common-sense approach has merit. If, in case after case, there is no weld metal failure under any test, the fault must lie with some other factor or factors. These could be adverse weather conditions, the wrong choice of filler metals, bad joint preparation, incorrect storage conditions of the electrodes, poor welding techniques, and plain incompetence of welders.

In Table 11-3, some of the meanings for the electrode numbers given in Table 11-2 are detailed. The first seven electrode numbers given in Table 11-3 are for the standard low-carbon steels. Since these steels comprise 90 to 95 percent of the steel used worldwide, their importance is evident. The first two digits of the numbers show the minimum strength of weld metal deposit. The third digit shows or indicates the position or positions where the electrodes can be used. The number 1

Table 11-2. Some factors to consider when choosing electrodes

AWS class	Tensile strength in P.S.I.	Yield point strength in P.S.I.	Elongation in 2″ (50.8 M.M.) by percentage
E-6010	(27.5 Metric Tons) 62000	(22.6 M.T.) 50000	22%
E-6011	(27.5 M.T.) 62000	(22.6 M.T.) 50000	22%
E-6012	(30.1 M.T.) 67000	(25 M.T.) 55000	17%
E-6013	(30.1 M.T.) 67000	(25 M.T.) 55000	17%
E-6014	(27.5 M.T.) 6200	(22.6 M.T.) 50000	17%
E-6020	(27.5 M.T.) 6200	(22.6 M.T.) 50000	25%
E-6027	(27.5 M.T.) 6200	(22.6 M.T.) 50000	25%
E-7010	(33 M.T.) (37.5 M.T.) 72000-81000	(28.5 M.T.) (32.0 M.T.) 63,000-71000	24%-32%
E-7014	(33 M.T.) 72000	(27.2 M.T.) 60000	17%
E-7015	(33 M.T.) 72000	(27.2 M.T.) 60000	22%
E-7016	(33 M.T.) 72000	(27.2 M.T.) 60000	22%
E-7018	(33 M.T.) 72000	(27.2 M.T.) 60000	22%
E-7024	(33 M.T.) 72000	(27.2 M.T.) 60000	17%
E-7028	(33 M.T.) 72000	(27.2 M.T.) 60000	22%

shows that it is an all-position rod. Some welders have mistakenly called it an all-purpose rod; of course, no such rod is available. The number 2 as a third digit indicates that it is for welding in the flat and horizontal positions only.

Table 11-3. Details about the digit system of electrode numbers

AWS Class	Electrode coatings for the welding of steel			
	Type of coating	Type of current	Welding position	Steels to be welded
E-6010	Cellulose sodium	D.C.E.P. Reverse polarity	All-for deepdigging	Low carbon share plate-pipe
E-6011	Cellulose potassium	Ac	All-for digging	Low carbon shades plate-pipe
E-6012	Rutile sodium	Ac-Fair best-ocen straight	All-best for flat horizontal	Low carbon shapes-plate
E-6013	Rutile potassium	Ac-best Fair D.C.R.P.-D.C.S.P.	All-best for flat horizontal	Low carbon shapes-plate
E-6014	Same as E-6013 with Iron powder	Ac-best Fair D.C.R.P.-D.C.S.P.	All-best for flat horizontal	Low carbon shapes-plate
E-6020	Iron oxide sodium	Ac best Fair D.C.S.P.-D.C.R.P.	Flat only or hog trough	Low carbon plate high deposition
E-6027	Iron oxide Sodium Iron powder 50%	Ac-best Fair D.C.S.P.-D.C.R.P.	Flat only Some horizontal	Low carbon plate or shapes
E-7010	Cellulose sodium	D.C.E.P. Reverse polarity	All- Deepdigging action	Low carbon shapes pipe-plate
E-7014	Same as E-6014 with better strength	AC best-fair for D.C.S.P.-D.C.R.P.	All-best for flat horizontal med. penetration	Low carbon plate
E-7015	Calcium carbonate Calcium flouride Sodium	D.C.E.P. Reverse polarity	All low hydrogen	Low alloy high strength shapes plate
E-7016	Same as E-7015 but with potassium	D.C.S.P.-D.C.R.P. or Ac	All low hydrogen	Same as E-7015
E7018	Same as E-7016 with some iron powder added	Ac or D.C.S.P.-D.C.R.P.	All-fair rather fluid	Same as E-7016
E-7024	Same as E-7014 but with more Iron powder	Ac or A.C.S.P.-D.C.R.P.?	Flat and horizontal	Low carbon shapes plate

Table 11-3. *(Continued)*

AWS Class	Type of coating	Type of current	Welding position	Steels to be welded
E-7028	Same as E-7018 but with 50% or more iron powder	Ac or A.C.S.P.-D.C.R.P.?	Flat and horizontal fillets	Low alloy high strength shape-plate
E-9016	Same as E-7016 with alloys added	D.C.R.P. or ac	All low hydrogen	Low alloy Higher strength shape-plate
E-11018	Same as E-7018 with alloys added	Ac or D.C.R.P. also D.C.S.P.	All low hydrogen	Low alloy Very high strength shapes-plate

The E-6020 was originally given a 6030 designation, which meant flat only. It had a trade name of Hot Rod. The puddle was so fluid that a 10-degree slope in any direction caused it to run rather than solidify in place.

The E-6010 and E-6011 electrodes are excellent for overhead welding. Common practice limits even these rods. The use of them in sizes larger than $\frac{3}{32}$ inch (4.7625 mm) is difficult if not impossible for the average welder. Most welding companies do not recommend using over $\frac{5}{32}$ inch (3.9687 mm) for the true overhead position. This is also true for E-6012, E-6013, and E-6014. The E-6010, E-6011, and E-7010 electrodes are often called fast freeze electrodes because the puddle solidifies almost as soon as the arc moves ahead.

The E-6027, E-7024, and E-7028 rods are called fast-fill electrodes. The deposition rates for these rods are very high. The E-7024, when run at amperages from 250 to 325, has a rate from 7.5 to 9 pounds (3.4 to 4.1 kilograms) per hour. The deposition rate for E-7028 run at amperages from 240 to 330 is 6.6 to 11 pounds (3 to 5 kilograms) per hour. The large-diameter electrodes in the E-xx14 to E-xx18 group are called fill-freeze. At current ratings from 215 to 290 amps, their metal deposit rate is 4.3 to 6.0 (1.95 to 2.75 kilograms) per hour. The E-6010 and E-6011 used at 135 to 200 amps have the lowest deposition rate, from 3.4 to 4.2 (1.56 to 1.91 kilograms) per hour. The rates for the E-6012 and E-6013 fall almost exactly between the last two groups. They have been called freeze-follow electrodes.

As seen in Table 11-3, six of the last nine E numbers fall in the low-hydrogen category. While other electrodes are available in each group, these figures provide most of the working information you will need. Please note the five-digit number at the bottom of the table. In the case of any five-digit number, the first three digits indicate tensile strength.

The last digit in any number has special meaning. That is why Table 11-3 is so important. While you may not care what the rod coat contains, you may need to have knowledge of columns two, three, and four. If you understand that the higher-strength numbers may not get all of their strength from the wire, that is fine. This means you must take care of the coat. Time, temperature, and moisture content are all vital factors in the transportation and storage of electrodes. You have no control of these things until the rods reach your company.

If you are doing work for or on a nuclear reactor, all materials require a detailed data compilation, which must include the following information: the exact location of the mine where the iron ore was obtained, the date it was dug, if it was stored before processing, how was it stored, how was it transported, what company ran the smelter, what process was used, what the chemical content of the steel was, when it was produced, if all the procedures followed, and if each person working with the process was able to sign off as it was completed.

The following questions must also be answered: Was the making of the wire done at that plant? If not, how was the steel transported and when? Was the content of the steel checked before the wire was extruded? What type of dies were used? Did each person involved sign for the date and hour? Was all the wire for this extrusion from the same batch of steel? Was any wire not used for these electrodes left over? If so, was it used for other electrodes? What process was used for obtaining electrode lengths? Were the cut wires examined for shape and grooving or ridging? Did quality control people sign for date and hour of inspection? How was the wire transported to the area for coating? The same process is required for each material that goes into the coat. This is especially true of alloys that may be added. Do you have a certificate of conformance to requirements for welding electrodes from the company? Was the certificate attested to by a notary public, and is the seal of that person on the certificate? It must use the company name and the date the material was supplied. The lot number and order numbers must be included. Did the electrode manufacturer show the joint configurations and the pass sequence for tests of this type of electrode? Was the test done this year? Did it meet AWSA # and ASME SFA #? Are radiographic results available?

Two-percent copper cuts the life of a reactor shell in half. The documentation of the finished electrodes continues: What type of packaging is used? Are containers sealed correctly? What type of transportation is used? Average temperature in transit was what? Are containers exposed? Are electrodes going directly to the worksite? If not, how are they stored? Is the work area protected from adverse weather conditions? Are the electrodes now kept at a controlled temperature?

Are they signed out to welders in a heated pack? Not more than 20 are given out at one time. Is the one-hour timetable observed? If a break is taken by the welder, the rods must be returned to be reconditioned. All rod stubs must be counted before any more are supplied. Of course, the materials to be joined have gone through the same process. Is the joint area protected? Have the correct codes been used for joint materials and their preparation? Have the welders been tested for these steels, their thickness, and joint design? Are the welders' qualification papers current? Do the welders know how to apply their identifying code stamps? Are the correct NDT (nondestructive test) procedures being used? Are the test results rechecked, signed for, and filed? Radiographic testing will show the number of passes and the thickness of metal deposited in each pass.

It has been said that when a reactor is complete, the weight of the paperwork equals that of the finished product. The exact steels you will be working with may not be known to any of the welding personnel, which could include the management people. The welding processes, the filler metals, and machine settings will be given. If the dial readings do not agree with the inspector's figures, all welding will stop. You may not be told, but metallurgical changes can occur at improper heat input and at interpass temperatures. This is very true of many of the stainless steels.

In some of these steels, particularly those with increased carbon content, the carbon changes. Carbide precipitation can make changes in the crystal structure of the steel. The carbides pick up chromium and move it to the grain boundaries. The crystal edges would then be very high in chrome, and the remainder has been depleted of chrome. The resistance of these chromium-nickel stainlesses to corrosion is greatly reduced. It can be somewhat restored by heat treatment. The high chrome-nickel steels are very susceptible. These austinitic stainlesses in the 14 to 18 percent chrome alloy content class are resistant to high temperatures and corrosion.

A long arc also presents problems in low-hydrogen welding. The difficulty here is the inclusion of hydrogen from the air around us.

The large companies involved in boiler construction, repair, and installation have had their own filler metals developed. These materials are for use in conventional and nuclear fired-pressure vessels. Since the erectors (companies) are totally liable for all materials and the work done by their employees, the fillers are excellent. These companies include but are not limited to B and W (Babcock and Wilcox), CBI (Chicago Bridge and Iron) and CE (Combustion Engineering). Their boxes or cans (thermetically sealed containers) have the company trade name on them.The superheater tubes on many large boilers carry live steam at 1200 psi (85 kilograms per square centimeter). Steam at these pressures has no odor, may have passed beyond the audial range, and is invisible. A pinhole leak would cut a person more quickly than a knife. The tubes are a chrome-moly alloy steel. The metal that has crossed the arc in the welding process is a very close match for the tubes.

The stamping of numbers on the stick electrodes does have short-comings. The numbers shown in Fig. 11-7 are clearly visible. Electrode sizes down to $\frac{1}{16}$ inch (1.5875 mm) and even $\frac{3}{32}$ inch (2.3812 mm) present a problem for stamping, so the old color codes are still used in some cases. Your local welding supply warehouse can get you hand-books detailing the filler metals available from the manufacturing companies. The cost will be minimal.

If required, welding cables may supply the current from the mains to the welding machines. They always supply the voltage and amperage from the machine to the following: the electrode holder, both stick electrodes and carbons; to the guns for GMAW and FCAW; to the contact tubes or tips for automatic submerged arc (SAW) and FCAW work. While the cables for your machines were probably supplied at the time of purchase, they may not be the best for you now. Remember, the ground cable should be the same size as the power cable. The cables (generally called leads) must be in good condition. The core of cable is usually of braided copper or stranded wires. A solid rod would carry the current but would not be flexible. The current is not carried inside the leads but is carried along the outside surface diameter of the copper. A good cable will have a light rubber wrap next to the copper. A flexible, open-weave wrap will cover it. The flex may have some insulating qualities. The final rubber wrap should be thick enough to handle whatever rough usage might occur. A small cut or scrape to the lead may cause real damage. If one strand of the copper wire is broken, it does not carry the proper current, which means there is resistance in the circuit. If old or cut cable coat

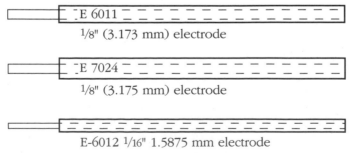

E 6011
1/8" (3.173 mm) electrode

E 7024
1/8" (3.175 mm) electrode

E-6012 1/16" 1.5875 mm electrode

11-7. *The electrode coat has an E-xxxx number for identification.*

is strung out across wet ground, the power loss will be very evident at the electrode.

This power loss is also always present over long distances with leads in perfect condition. Think of the welding machine as a water pump. At a distance of 10 feet from the pump, an excellent force propels the water. At 200 feet (approximately 61 meters), that force is greatly diminished. If the pipe has leaks, there may be little water available where it is needed.

Some home welding units may be purchased with #4 or #5 cable. If your machines are for industrial use, always use #1/0 cable. If you are using 300 or more amps, try #2/0, especially if you use 100 feet (approximately 30.5 meters) of cable. As the amperage usage goes up, so must the cable size. There are #3/0 and #4/0 cables readily available. At 500 amps you may need the #4/0 for even short leads.

If you must weld 300 feet from the power source, start with a #4/0 cable, then go down to #3/0. Then go down to #2/0 and finally use #1/0 at the electrode holder. This method works as well as necking down pipe sizes and putting a nozzle at the end. For using long cables, make sure that the fittings are the proper size and make a tight-locking connection.

When you are welding with fully automatic equipment, you may need two or three #4/0 leads, especially if you are using an amperage range from 800 to 1500. The duty cycle of a machine is always a factor in your cable selection. In all machines for automatics, they should have a 100-percent duty cycle, a constant-voltage, constant-amperage welder. If your machines have a slope control, you might experiment a little, but very little. Beyond 450 amps, slope may damage the machine if cranked in for several turns.

In the distant past, welding machines were sold on how they performed at that moment. Now they are sold on how they perform as rated. Some machines had a duty cycle of 20 percent. A 180-amp welder would run one $\frac{3}{16}$-inch (4.7625-mm) E-6012 electrode. Then you had to wait eight minutes before welding with a second one. The NEMA rating was not always observed. That rating allows you a small margin for running at capacity without damaging the machine.

Duty cycle means simply the number of minutes out of any 10 minutes that the machine can deliver the same amount of current to the arc. A 60-percent duty cycle welder lets you weld for six minutes out of 10 at the full rated capacity. If you must slag a weld, remove a rod stub, and replace it with a new electrode, you can probably weld at full rating without damaging the machine. If you need a 100-percent duty cycle machine, don't settle for less.

There are two other welding processes that you may run into. The first is electroslag welding. It was truly improved by the Russians in their testing facilities. It is perfect for welding thick plates and castings. Little or no edge preparation is necessary. Material up to 30 inches (77.2 cm) thick can be welded without difficulty (Fig. 11-8). This type of welding is almost a resmelting or casting process. A backing plate is placed under the square butt joint. A layer of flux covers the backing at the bottom of the joint. The arc initiates the process but is damped out by new material. The current continues to pass through the molten metal, slag, and flux. The resistance now provides the heat until the weld is completed.

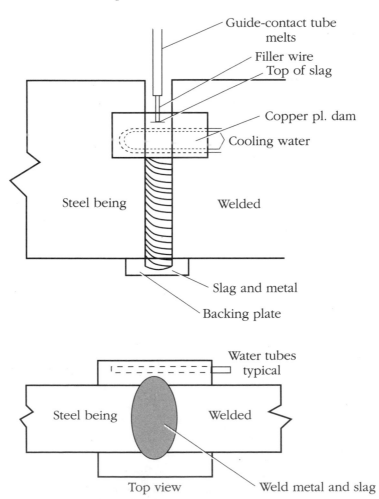

11-8. *The positioning of steel and the type of wire gun for the electroslag process.*

The square butt joint opening is from 1 inch (25.4 mm) to 1½ inches (38.1 mm) wide. The thickness of plate dictates the number of filler wires needed. A 3-inch (76.2-mm) plate uses one wire. A 26-inch (67-cm) joint can use five wires and power sources. The deposition rate is quite high, because each wire may deposit up to 35 pounds per hour. The dams or shoes that move up the joint vertically as the joint fills are made of copper. The weld metal does not stick to it, and it is a good heat conductor. The dams are water-cooled. They usually have an area for the slag and a little weld metal to protrude from the joint edges, which takes care of complete fill and metal shrinkage while cooling. The wires provide better coverage if a weave pattern is followed.

The second process is the electrogas process. It is very similar to electroslag welding. The position is vertical. The joint configuration is the same. This process is a true arc process. The filler is cored wire. The other difference is that outside shielding gas must cover the weld. The gas and cored wire used are dictated by the metal to be welded. This is calculated by the thickness of the plate or castings to be joined. This process will become more popular as cored wires improve and the floating of slag also increases.

As promised, the further study of stick electrode welding problems will now be addressed. While coated electrodes have greatly reduced the problem of arc blow, they have not completely solved the problem. In one case, arc blow is really a magnetic current within the work and the flux on the rod. In this case, the arc blow is away from the ground. To a great extent, this problem can be at least partially solved by connecting two ground cables to the ground tap on the machine. Then clamp them to the work on both sides of the joint and travel toward them. If this is not effective, try moving the grounds to different positions on the work. If extreme magnetic arc blow is present, it will be almost impossible to fill the crater at the top of a butt joint in the vertical position.

A problem condition can exist in your building. If copper pipes are laid in the flooring or the piping is housed in conduits in tunnels beneath the building, problems will persist. You may solve part of the problem with standard arc blow by simply angling the electrodes to compensate for the field.

In the case of test plates, the backing plate behind the open butt joint can be extended beyond the joint and welded to both plates at each end. In the case of deep groove or deep open butt joints, the problem will increase. As you might suspect, you will be asked to increase your rod size, and this increase always increases arc blow. You can try wrapping several turns of ground cable around the work.

Now for a better answer: Change to ac power. The culprit is often dc power.

Electrode manipulation

You must run the weld bead, it cannot run you! If we choose the electrode, the machine, machine settings, and joint formation, and they are relatively correct, then we have a good starting point to make an adequate weld deposit. We must also use knowledge of the following things: parent metal, its carbon content, any alloys, position for welding, and thickness.

The next item to consider has been mentioned, but let's take a closer look at electrode size, expected elongation factors, tensile strength of the weld deposit, and the coat makeup (gas and slag formers, any alloys, and percentage of iron powder). Also consider angles of incline and lead. (Remember that incline is to the right or left of direction of travel while the lead is relative to and in direction of travel.) Even that factor (direction of travel—up, down, forward) and bead size must enter the picture.

The electrode choice dictates the depth of penetration and bead contour as well as ease of slag removal. E-7028 will suck hydrogen into the deposit if the lead angle is such that it comes closer than 60 degrees to the plate in the direction of travel.

Deep groove and bevel butt welds with E-7024 will produce an intense slag removal problem. In fact, the Lincoln Electric Company has produced three different electrodes in this class as jet weld 1, 2, and 3 for specific slag problems.

Most welds made in the vertical down-hand mode do not penetrate deeply into the parent metal. The welds are not as wide as those using a weave or wash pass and welding vertically up. A slight weave is acceptable, but fluid slag prevents a true U or triangle move. It is fast, and most pipelines utilize the technique. Remember the line wall thickness never exceeds ½ inch. The rod angle for pipelining is angled sharply upward from the classic up-hand point toward the theoretical pipe center. Even with ³⁄₁₆-inch E-8010, the actual increase in weld thickness never exceeds ³⁄₁₆ inch (4.7625 mm) and usually is limited to ⅛ inch (3.175 mm) on any one pass. The same weld made up-hand can carry ⅜ inch (10.375 mm) of metal, affect a deeper tie to the parent metal, and result in a much wider bead face. The burn-off rate of any electrode is simply the product of the amperage at the arc and the cross-sectional area of the electrode. (The more amperage in the circuit, the more rod melted and the greater deposition efficiency.)

The directing of arc force enables us to weld efficiently in horizontal and overhead positions. The hook or J oscillation move allows us to push metal onto the upper plate surfaces, and the same manipulative moves along with the timed holding of the arc at the back of the puddle effectively corrects for the undercutting action of the heat and force of the arc.

The near perfect spacing of ripples on the surface of your welds are proof of your mastery of manipulative skills. Each gas bubble pushed to the surface and each slag pocket filled and that slag moved to its proper place at the top and back of the puddle tell you that it's your move.

Fingernailing

The condition known as fingernailing is intolerable when absolute weld deposit quality must be maintained. Fingernailing is a condition in which the welding electrode coat burns away more on one side of the rod than the other. If we examine all the factors that could contribute to this problem, we can probably eliminate some of them.

The first solvable factor is uneven moisture content on one side of the electrode coat. The cause might be improper drying or packaging where moisture could impinge on one side. Green or improperly cured rods may actually stick together in blocks. Although a rod drying oven will not solve the green-rod situation, it will if electrodes are rolled on the racks of the oven, evening the moisture content on each side of the coat. One-side color change is always suspect in any electrode examination. If rods show color change from one sealed container to another, moisture content and drying time or temperature are the reasons.

Too little moisture will cause the coat to crack or revert to powder and flake away from the wire. In the case of E-6010, E-6011, E-6012, and E-6013, loss of moisture beyond the recommended level may be disastrous. Any cracks are an invitation to arc blow out. Powdering will lead to uneven coat thickness and to fingernailing.

In most instances, rod diameter is small, not over $5/32$ inch. The smaller the electrode, the more often we can expect the problem to occur. If fingernailing is evident, of course, the vertical and perhaps the overhead positions will provide the real problems.

Try changing the electrode in the holder a full 180 degrees. Doing so will prove uneven rod coat or uneven moisture content. Another factor could be the wire itself. Wire cast, uneven wire thickness, or curves and kinks will produce the condition.

When checking one-side coat thickness, remove the coat from one side down to bare wire. With a micrometer, measure the coat and the wire, but do not use undue pressure on the coat. Check the other side in a like manner, a short distance up or down the rod. The variance should not exceed 0.002 or 0.003 of an inch. Check each electrode prior to use in critical situations. Coat lumps, bare spots, cracks, powdering, and spot discoloration are reasons for discard.

Operator techniques are often at fault in fingernailing. Too low a current, improper rod angle, and poor joint balance are the greatest problems.

Fingernailing on the root pass of a butt joint will cause instant failure. If it is a welding test, you may not get a second chance. Inspect each rod very carefully before starting to weld.

When faced with welding rusted or galvanized steel, the first tendency is to say, "I won't do it." Still, at some time you will be faced with the task of doing both. Galvanized steel has been dipped in liquid zinc. Zinc, when heated, gives off zinc oxide, which is a deadly and cumulative poison. In fact, any heavy metal, when taken into the body, stays with you, at least in a large percentage. If you fusion-weld galvanized material, remember, steel and zinc do not mix. The zinc should be removed by grinding, chipping, or machining. You must use respirators or masks for this job and change filters often. Any residual zinc retained at the surface is easily burned off with the first pass. I would recommend E-6010, E-7010, or E-8010 electrodes for the job, because of their deep penetrating action and their ability to *float* impurities to the top of a bead. E-6011 would be my next choice. Wire and even other stick electrodes will not do an acceptable job.

Rust does not produce toxic fumes, but descaling (rust removal) certainly requires a dust mask. If prior rust removal is not possible, again, E-xx10 electrodes are recommended. Always work from solid metal into rusted areas. The arc action is similar to that of working from heavy plate to sheet metal. The tie into the solid metal establishes the puddle, the hooking or pushing movement into the rust and the backward oscillation pushes slag and forms the ripples.

Remember, the rust on steel is in plates or scales. The breaks in the scale permit additional oxygen to reach the surface and the rusting process continues. The rust on nonferrous metals leaves a continuous coat, so after the coat thickness is established, no further oxidation occurs. When welding nonferrous metals, oxides should be removed prior to welding operations. The reverse-polarity half of the sine wave generated using the GTAW process also breaks up nonferrous oxides, and the gas shielding does not allow oxygen to attack the metal while it is still hot. Hot metal is most susceptible to oxidation regardless of type.

Backing plates for tests are always mild steel unless otherwise specified. The plates on strips are from ⅛ to ⅜ inch thick and wide enough to allow for some heat dissipation factors. There should be no melt or burn-through on backing plates. However, complete fusion must be obtained. Care must be taken that space does not exist between parent materials and backing plates that are not to be removed, as slag will move into the space and will show up as flaws on X-rays or ultrasonic scans.

While we are primarily concerned with stick electrode processes, it is wise to remember that GMAW and SAW use very high voltage and amperage ratings. Any increase in these factors in relation to the cross-sectional area of the filler metal will result in greater depth of fusion with parent and backing materials.

Inserts come in many forms. One of the most commonly used is the backing or *chill ring* for pipeline applications. While one variety of insert fills and spaces the root gap opening with filler material, another has only spacer lugs and a backing ring. With this last ring, tacks must be made in the root, tying the pipes together with sufficient melt-through to maintain complete fusion with the backing ring. The spacer lugs, usually ⅛ inch in diameter or less, are then broken off and you are ready to *feathergrind* the tacks and complete the root pass.

Backing rings allow greater amp-volt settings and, of course, a greater controlled melt-through and complete deep penetration into the ring. If care is not taken, the increased current ratios may result in undercutting one or both sides of the bevel, and slag may be trapped by the subsequent hot pass.

Patented materials are available as a backing tape that is sealed to one length of pipe and extends far enough in width to cover the root opening as the next joint is moved carefully into place. The tape is used mainly for GTAW processes. The insulating materials are covered by aluminum foil, which would attract oxygen as it burned. Of course, this is precluded by your gas coverage. The purpose of the tape is to prevent gas escaping into the pipe, and thus it shields both the top and bottom of the root pass. For test purposes, the backing is removed, usually by machining, and testing is done in the usual manner.

Other materials, such as copper, are often used as backing plates. The copper is grooved to provide a space for melt-through. While copper is a great heat conductor, that ability is often decreased by the circulation of water flowing through passages in it. Since copper does not alloy itself readily, it does not mix with or adhere to the parent or filler metal as the welding process takes place.

Carbon is good backing material. Block or slab forms are very useful. Any smelter that uses arc furnaces should have short stub electrodes for sale. To be effective, it should be about ⅜ inch thick. Carbon in this form is easily shaped using any common woodworking tools.

A good welding supply house carries stick carbon in flats, strips, or round dowel shapes. If a threaded hole in a part to be repaired must be saved, a dowel of the proper size (external thread size) can be screwed into the hole with no more difficulty than putting a bolt in with finger-hand strength. Remember, carbon will precipitate into parent or filler metal up to the amount found in that particular steel and will be at the upper range for that type of steel.

Project 9: Truck bumper

This bumper can be fitted to all pickups and light trucks. It will withstand impact values of at least two to three times that of stock bumpers.

The 5-inch channel should be 7 feet in length. Of course, if you wish to use scrap, it can be spliced in two or more places. Do not put a splice at any cut or bend line. The diamond plate can also be spliced as needed. If these materials are welded and ground correctly, they will not show joint marks after they are painted or chromed. The brackets can be mild steel (low carbon), or, if scrap low-alloy, high-tensile-strength steel is your choice, then $\frac{3}{16}$- or $\frac{1}{4}$-inch thickness is adequate. The alternate style brackets may be used if the bumper should need to be lower to fit a specific trailer.

The flanges of the channel should be cut at points A and B in Fig. 11-9 and also at the exact opposite points of the hitch plate. The web of the channel should be heated or bent in a press until the proper angle is obtained. You may want a cardboard pattern to make both bends the same. This will leave a triangular opening at A. You will need to remove a small piece of each flange as shown in detail B. These pieces can be inserted at point A. Bend the channel web back until it is again parallel with the license plate area. Use the same bend method.

Find the distance required to have 1 inch of clearance beyond the fenders. Make the cut-out as shown in detail C. You need to heat these bends to form a nice radius contour. A dull red color, 1200 to 1400 degrees F, is fine. You will now need to decide how much fender protection you need. If the distance from C is more than 10 inches, you may want to add one or two gussets there and on the opposite end of the bumper. If your channels are in line and you are satisfied with the overall appearance, add the diamond plate as shown.

Fit the bumper to the truck. Is the clearance correct? Find out what style of brackets you need. Clamp the brackets to the truck frame and find out where they fit the channels of the bumper. You may want to tack-weld them to the channels. Mark the holes in the truck frame to the brackets. A fine-pointed marking pencil or soapstone will be adequate. The holes in the truck frames are $\frac{1}{2}$ inch in

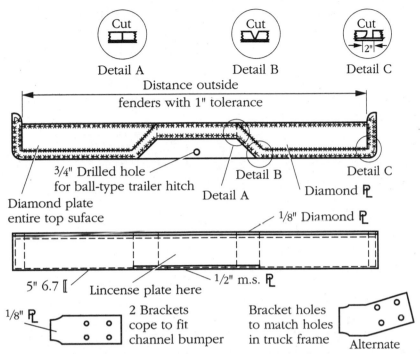

11-9. *The fabrication and welding of a truck rear bumper.*

diameter. Match them as closely as possible. Center-punch the hole centers on the brackets. Drill or punch the holes using the iron worker. You can break the tacks from the channel if necessary for this process. Refit the brackets to the frame. If you need to ream (enlarge the holes), be very careful. Redrilling holes, especially in tight places, may cause the drill bit to hang up and twist the drill motor and cause your hands to hit the body or frame of the truck.

For strength and durability, you may want to use treated cap screws rather than bolts. When the bumper is finished, fit it one more time to the truck before prime painting or sending it off to be chromed. Since it will be subject to road abrasives and perhaps salt spray, you should use a zinc-oxide primer. You can match the paint trim or use metallic silver or aluminum spray paint for a top coat. Use respirators and all other safety devices when spray painting.

Questions for study

1. How does bonding differ from fusion welding?
2. What actually produced the first arc action?
3. Could you ever use bare welding rods? If so, when?
4. What was used to make the coat on barber-pole electrodes?

5. Who gets the credit for our present electrode numbering system?
6. When four digits are stamped on a rod, what do the first two stand for? The third? And the fourth?
7. Would stainless steel be your first choice when making a repair on Hadfield-grade, austinetic manganese? If not, why?
8. What are green rods?
9. Where would one use the submerged arc process in welding?
10. What is the arc air process?
11. What does GMAW stand for?
12. In what way is water detrimental to a weld puddle?
13. What is skelp?
14. How is GMAW different from FCAW?
15. How does a shade #5 differ from a shade #12 dark lens?
16. What is a flash burn?
17. In what way is the use of large wire packs an advantage in FCAW?
18. How does spalling occur?
19. What is galling?
20. How do you sharpen a thoriated tungsten?
21. What happens if the tungsten touches the work in GTAW?
22. What is a hot start?
23. Why does E-7028 have a higher deposition rate than E-7018?
24. How does the elongation factor affect the weld metal?
25. What is an all-purpose rod?
26. Give the E numbers for two electrodes with a low deposition rate.
27. Why is a five-digit E number necessary?
28. Can you name 10 things that must be documented before electrodes can be used on a nuclear reactor shell?
29. How will a long arc affect the welds made with LH (low-hydrogen) electrodes?
30. Why are color codes still being used on some electrodes?
31. What is the makeup of a welding cable?
32. When you are using 500 amps, what size cable should be used?
33. How much slope would one use when welding with 500 amps?
34. If electroslag welding is not a true arc process, what is it?
35. What happens to weld metal as it cools?
36. Can you name three things that will help minimize arc blow?

37. List three factors that can cause fingernailing.

38. What E-number electrodes are a good choice for welding previously galvanized steel?

39. What kind of poisoning can result from welding galvanized steel without proper metal preparation?

40. Can you suggest a type of oscillation that might help in the welding of rusted metal?

12

Testing and codes

Do not think of testing as a function of a welding process only. A good example of this has been previously mentioned. Bridges in the United States that are at least 15 years old are in need of repair or replacement. While some major types of weld problems have been examined, the areas of riveted and bolted construction have also shown fatigue cracking. Some of the cracking could be simple design mistakes. Poor maintenance is even more of a factor. The bridges in our local area have loose and broken rivets. Since many of these bridges are on secondary roads, federal inspection is not required.

What was the cause of fastener failure? Any one of the stress factors noted in Chapter 10 could be the culprit. As the bolt or rivet snapped, scarring or grooving of the steel around the hole would weaken the area. Not one of these bridges have completely failed, which tells you that the engineering was excellent for the time period involved.

Most steel now used in these structures not only follows ASTM code but is often examined by ultrasonic testing procedures. This method of testing is explained later in this chapter. Cracking in weld areas has been studied extensively. For the most part, these studies have involved large bridges in the eastern United States. To be truly conclusive, the examinations would have to cover both time and temperature conditions. Alaska and Texas would be good test areas.

Because of the age of some structures, different weld processes had to be involved. The earliest involved SMAW, and the latest could have been welded with FCAW processes. In between could be GMAW and even electrogas or electroslag welding.

If you can eliminate the places where weld problems were designed into the bridges, then we have personnel problems. The welders did a poor job, and inspectors failed to do their job. You don't need figures to tell you that pervasive hydrogen caused cracking. All welding should be protected. Rain, sleet, and snow may not stop the postal service, but it should stop the welding process.

The care of the HAZ (heat-affected zone) was probably a factor. On thick sections, were preheat recommendations followed? Was lack of post-heat treatment a factor? Were the welders qualified to do the job? Were samples of their welding proficiency tested before they were allowed to start a weld on the job? Melt-through (root penetration) may have been inadequate. What kind of inclusions were present? Was there evidence of inline porosity? If there was crowning of weld in deep grooves, was there enough voltage to dig into the parent metal at the sides of the crown?

Figure 12-1 shows *wagon tracks* (incomplete fusion with both parent metal and previous weld passes). For lack of root fusion, there should have been back grooving to sound weld metal. The crown should have been ground to prevent the wagon tracks. These same types of problems will be present in any fabricated item.

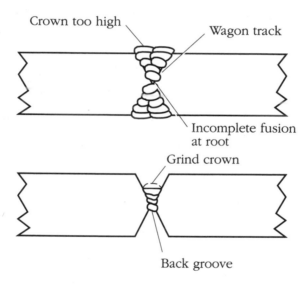

12-1 *The high crown of a weld bead and the possibility of second-pass incomplete fusion (wagon tracks).*

In many cases, the prime contractor will require the testing of all materials used and further testing of the welding personnel. Finally, the joints must follow set rules, and usually these rules must be documented. The weld themselves will often be subject to NDT (nondestructive testing). The HAZ is not always subject to testing.

The ASTM has set many of the standards for the steels used. These include ship steels (hulls in ships), structural steels (carbon steel plates and structural quality), steel for structural shapes, plates, and bars, structural steel for locomotives and cars, steel for welded structures, steel for bridges and buildings, carbon-silicon steel plates for machine parts and general construction, standard low-and intermediate-tensile-strength carbon plates of flange and firebox qualities, carbon-silicon plates for intermediate tensile ranges for fusion-welded boilers and pressure vessels, high-tensile-strength carbon-silicon plates for boilers and pressure vessels, high-tensile-strength carbon-manganese steel plates for boilers and pressure vessels, and high-tensile-strength carbon-manganese steel plates for unfired pressure vessels. The ASME has also placed and required similar tests in seven of the 13 categories of steel just listed.

The American Bureau of Shipping Rules (ABS) has test standards for hulls and ships. The United States Coast Guard (USCG) have tests for marine boiler steel plates, and the Association of American Railroads (AAR) has tests for locomotives and rail cars.

The tests for steel plates and sections may include mechanical tests but no chemical requirement limits on any element. Where chemical requirements are needed (carbon-manganese steels), the mechanical tests are still required. There are cases of limits on one or more elements with no mechanical test required. If you need steels of a specific tensile strength, check with ASTM.

If you have a specification such as A-370 for your job, you may be asked for a small fee to cover costs of current data. Actually, the chemical composition is a prime factor. This includes carbon content and treatment. ASTM can also supply elongation and bend test requirements. The A-370 test data will probably include results of testing on steel with your A number.

In Chapter 1, the Tinious Olsen testing device is shown in Fig. 1-9. ASTM will supply data for the stress-strain curve, the elastic limit, and the ultimate tensile strength of your steel. The Brinell hardness factor will also be given. These tests protect you and your company in case of litigation on product liability.

The Charpy impact test involves a steel specimen. The steel is of a given size. It is notched and placed in a vise. The notch is placed just above vise jaws. A hammer-type device is mounted on a pendulum. The impact force is calibrated as to length of swing and hammer weight. The results are stated as impact strength. These strengths can be given at room temperature or as low as -40 degrees F (approximately -40 degrees C). You should understand that embrittlement will occur as temperature decreases.

Macroetch tests are done on steel specimens dipped in hot acid solutions and examined under a metallurgical microscope. This type of test is not generally done on steels for standard use. It is real help to investigators of metal flaws or failures in aircraft accidents. Ships and submarine problems are also examined in this manner.

Magnetic particle inspection can be done on steel sections or plate edges. While this test is more often used on welded materials, it will spot laminations and other discontinuities. The plates or shapes must first be magnetized. A powder of ferromagnetic grains is spread over the test area. The particles will show heavy concentration where any problem exists on or near the surface of the steel. This test can be used in weld processes for each pass as the weld material is deposited. This test can be used on most steels. The exception would be some stainlesses that are nonmagnetic. It will show incomplete fusion and unrepaired undercut. It does not require the surface cleaning noted with dye-penetrant tests.

The coil method of wrapping steel or a yoke that can be moved on suspected areas is fine where the current is applied. Even after the current stops flowing, the steel will remain magnetized. Most powder for such testing is gray in color. If a liquid is used as an agent for the particles, fluorescent materials may be carried in the liquid. They are set to glow in the dark with black light. Dc current is best for demagnetizing the steel, but heating above the temperature where it becomes nonmagnetic is, of course, the final answer.

We do not think about the most common test. When you look at steel, what do you see? Is the steel exactly what you expected when you first saw it? Visual inspection is important and present at every part of the fabrication welding and testing process. When the steel arrives at the work place, you check for size and shape. As you become familiar with the steel commonly used by your company, you will spot changes in your material almost instantly. The thicknesses of plate and flanges on shapes are indicators. The ends of beams and channels will show web thickness. These factors will give you a check on weight. Are there plate imperfections? Check for steel not completely a continuous part of the surface. Are there laminations at the plate edges? What about rusted or pitted surfaces? Does the steel appear to be the thickness ordered? If in doubt, check with a gauge, caliper, or micrometer.

Remember that some tolerance is allowed on all shapes and plates. You will need to see that the flanges on the shapes are close enough to parallel to serve your requirements.

The letters VT always stand for visual testing (inspection). If this designation appears on quality control documents, you must establish a standard checklist. The list is critical for materials, but many welds are not considered critical. They should be noted as visually inspected. Some responsible person should sign and date each item on the list. If an item is NA (not applicable), sign as such.

At the present time, there are no societies or recognized institutes that offer licensing of steel fabricators. The boiler and pressure vessel departments of many states do have licensing requirements (Fig. 12-2). You will notice that the first card states the type of work you are licensed to do. As the years pass, the processes are no longer spelled out. Finally you have a *class* of license.

If you are working in the trade as a structural steel fabricator, and such licensing is offered, you may be fortunate. Some testing may be required as is the case for boilermaking. Some people who could show proof of years of work on boilers applied under a grandfather clause. To start the new licensing of a trade usually recognizes that certain persons are well qualified. If such a license is offered, you should accept immediately. If minor critical areas of welding are going to be required by your company, you might be considered for the job as a checker of welds and procedures.

The AWS offers courses towards licensing of inspectors. The AWS standards are high and you will be tested before any license is granted. The three grades of inspection licenses cover all of the fields from visual through radiographic qualifications. For the radiographic license, you must study under a person holding such license for a period of one year. In part, the health and safety requirements make this essential.

If you or your company must document a welding procedure for a fabricated and welded item, you will need the following: a detailed method of welding each joint, the materials involved, the practices and equipment involved, the type of process, the filler metal or metals used, the shielding gas or gases, if required.

If you are using a manual or semiautomatic process, you are the welder. If it is automatic or machine, you are the operator. The base metal specimen will be supplied by the company. This is very true if different ASTM numbers are involved. The same is true for other material numbers, fuel gas, if any, and filler metal class. The use of any backing plates, tape, or inserts must be noted. In GTAW, extra gas may be supplied to the back of a groove weld. You will set the current.

Is the weld a double-groove weld? Since you or your company will want you to be qualified for as many weld situations as possible, you can make the test cover a variety of code requirements. The groove test covers the fillet weld qualifications. The double groove covers single groove. The short circuiting is GMAW or submerged arc processes.

BUREAU OF LABOR
STATE OF OREGON

September 11, 1957
Date of Test

This is to certify that .... John W. Shuster

is qualified to perform .... Manual Arc

welding in all positions with or without a backing ring on valves
pipes and fittings installed beyond the main and/ or auxiliary
stop valves.

(Subject to provisions on reverse side)

Chief Boiler   Commissioner
Inspector

BUREAU OF LABOR
STATE OF OREGON

March 4, 1970
Date of Test

This is to certify that .... JOHN SHUSTER

is qualified to perform .... TIG

welding in all positions with or without a backing ring on
valves, pipes and fittings installed beyond the main and/ or
auxiliary stop valves. **Max. thickness 5/8" P-1 Ma**

(Subject to provisions on reverse side)

Chief Boiler   Commissioner
Inspector

STATE OF OREGON

BUILDING CODES AGENCY

BOILER AND PRESSURE VESSEL SECTION
This certifies that the person named hereon is licensed/
registered as provided by law as a

CLASS FOUR

License no.: E76 - 1995

Expires: 07/31/96

SHUSTER JOHN W

Signature of licensee

DEPARTMENT OF
CONSUMER
& BUSINESS
& SERVICES

12-2 *Boilermaker licensing.*

STATE OF OREGON

BOILER AND PRESSURE VESSEL SECTION

REG./LIC.AS: No: E76-1995

CLASS FOUR

SHUSTER JOHN W

                                            EXPIRES:

BUILDING CODES AGENCY   07/31/93

STATE OF OREGON

BUILDING CODES (BOILER)

REG./LIC.AS: No: E76-1995

BOILER MAKER MECHANIC

SHUSTER JOHN W

                                            EXPIRES:

DEPARTMENT OF COMMERCE   07/31/88

BOILERMAKER MECHANIC FOR A
LICENSED INSTALLER,   No.
ALTERER OR REPAIRER

STATE OF OREGON

DEPARTMENT OF COMMERCE

John W. Shuster

ISSUE DATE   EXPIRATION DATE
7/8/76   7/8/77

12-2 *(continued)*

The tests are generally on pipe or plate. Rebar is not usually butt welded. The welding positions are 1G flat, 2G horizontal, 3G vertical, 4G overhead. 5G is a pipe test with the pipe in horizontal position. 6G is also a pipe designation. The two sections of pipe are beveled regardless of pipe diameter. In any case, the plates or pipes are each about 5 inches (127 mm) in length. For the 6G pipe test, the pieces are tacked together with a set root opening. The material is set on a 45-degree angle to the horizontal plane. With this test, there is no true flat-position welding involved. If you pass the test without backing, you are also qualified for welds with backing and for 1G, 2G, 3G, and 4G.

Many companies also consider it a qualification for groove plate tests. The groove tests are often used to supersede the fillet tests. The plate or pipe thicknesses are another factor. Schedule 120 (triple-strength pipe) qualifies for schedule 40 and 80.

A test on plate can involve a 1-inch (25.4-mm) thickness. This test qualifies the welder for materials of the same A number to unlimited thickness, except sheet metal.

Rebar (reinforcing rod) may be tensile or bend-tested just as most plate and pipe. Of the six tests given, the visual and bend tests are most common. In general, if an inspector sees flaws in test specimens, he or she looks the test out without even completing any further testing.

For the steels beyond ¾ inch (19.050 mm) in thickness, a side-bend test is done. The standard mandrel as shown in Fig. 12-5 is used in all such tests. Figures 12-3 and 12-4 show the procedure for taking test coupons from ⅜-inch (9.525-mm) and 1-inch (25.4-mm) plate.

It should be noted that ⅜-inch (9.525-mm) plate test qualifies up to the ¾-inch thickness unless otherwise stated. If a test specimen fails, it is almost always at the root. Of course the side-bend tests may fail at some other point, but the root is still the critical point. This root area is also the point of most concern in pipe tests.

The problem with the side bend is worth discussion. The filler metal is often of higher strength than the parent metal. When tested, this metal resists bending and bulges slightly. The fusion zone is subject to unequal stress patterns. If any slag inclusions or porosity is present, it will show on the radius of the bend. For American Petroleum Institute (API) and some AWS tests, a small surface blemish or dimpled area may appear on the top of the radius. In these cases, the flaw should be ¹⁄₁₆ inch (1.5875 mm) in diameter. The dimple indicates a tiny subsurface area that has collapsed. Some API tests are even more lenient.

The 5G test is for pipe welded vertically down. The tests are the same for DT (destructive tests) to qualify welding personnel. Most pipe and tank work involves some government dollars.

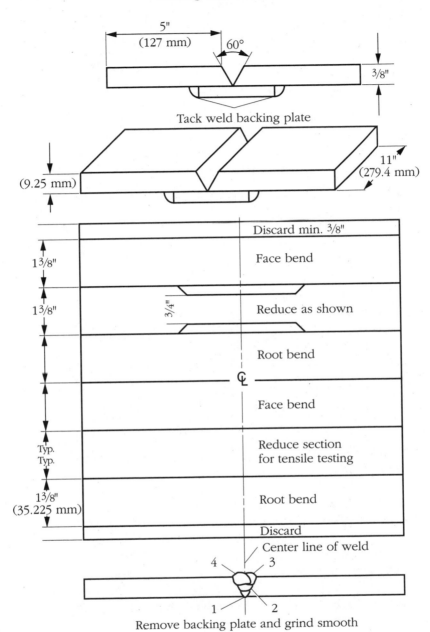

12-3 *Location of test specimens for welder qualification.*

In many cases, a radiographic film record is kept to show the integrity of the welds. The vertical down-hand welding of pipe is much faster than the vertical up-hand welding procedure. In welding vertically down, the root pass must use a small, even-size keyhole to

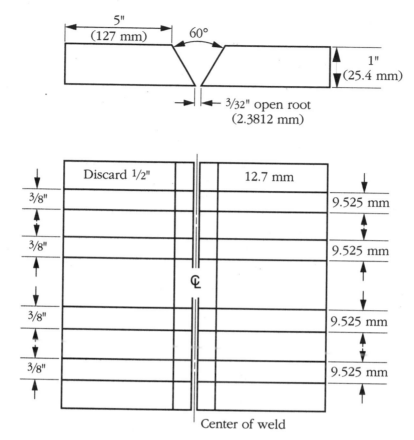

12-4 *The placement of coupons and section removal for side-bend specimens from thick plate.*

ensure complete melt-through (Fig. 12-6). The size keyhole and wall thickness of the pipe dictate the size of the weld deposit. If too large a keyhole is formed, excessive burn-through will occur.

Practice on plate of like thickness will soon show you the correct amount of metal to carry and the right current settings for *you*. In a perfect procedure, you start at the top of the pipe; always aim at the theoretical center of the pipe. As the weld draws on the pipe, the root gap may grow smaller. For good current settings and SMAW processes, the size electrode dictates these factors. In any case, 25 to 26 volts should be considered an upper limit. If the root gap is too small, the electrode angle can be changed to a more upright position for deeper penetration.

If the root gap should increase due to poor fit or the current setting being too high, change the rod angle to one more parallel to the

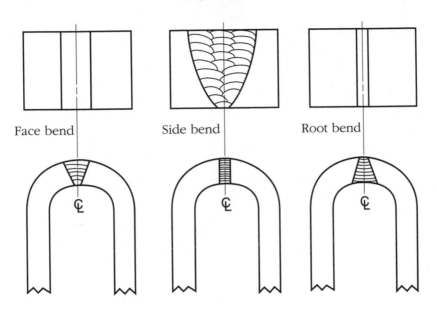

Face bend   Side bend   Root bend

12-5 *The use of a standard mandrel on root and side bends.*

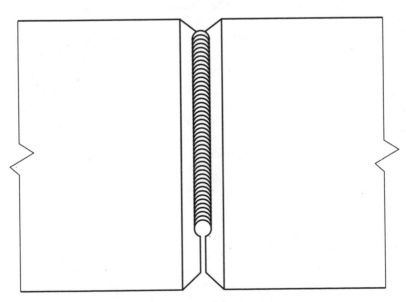

12-6 *Keyhole size for welding vertically down on pipe.*

top and bottom surfaces of the pipe to decrease penetration. Beware of electrode angle change when using low-hydrogen filler metals. Never let the slag move into or ahead of the puddle.

The very proficient pipeliner can use a small root gap and bury the arc. Listen for the sound of arc as its force is now inside the pipe. Practice until you can see a uniform melt-through and know you are identifying the correct sound.

You may also want to experiment with a complete bevel or a chamfer and a *land*. If the bevel leaves too sharp an edge, grind the knife edge to a uniform root face (land) with a power sander or grinder. This land need not be more than $\frac{1}{16}$ inch (1.5875 mm). Don't vary the land size.

With practice, you can carry 125 to 170 amps using E-7010 electrodes and $\frac{5}{32}$-inch (3.9687-mm) diameter on mild steel pipe. The wall thickness for these factors may range from $\frac{1}{4}$ to $\frac{1}{2}$ inch (6.35 to 12.70 mm). The $\frac{1}{2}$-inch and thicker wall pipes may require up-hand welding, which will be stated in procedure documents.

In any case, the welding of pipe can require good welders and a good inspection team. Just as a doctor who specializes in reading X-rays is often called in to verify conditions, you need a specialist. The protective slag on the inside of the pipe on the root pass may be misinterpreted as internal undercut. Several other types, such as heavy ripples may lead to the same false conclusion. The *wagon tracks* previously mentioned are real problems. They will usually appear as two dark lines of varied length on both sides of the centerline in a good radiograph. If it shows up on one side of center only, it could lead to a false conclusion. One thing that can produce wagon tracks is a narrow root opening. If you find the included angle of the bevel is too small, again use the blade or disc of a sander or grinder to enlarge the opening.

Pipe in enclosed areas may not have this last problem, but pipelines and penstocks do. If adverse weather conditions exist, you should preheat. At least bring the steel up to 70 degrees F (21 degrees C), and use wind breaks as needed.

The qualifying test specimens for DT should be taken from the welded sections as shown in Fig. 12-7. As noted, all of the coupons are taken from the pipe or tube at positions that favor the welder. The possible exception is the vertical root bend specimen.

Some procedure qualifications may call for the coupons to be 1 $\frac{3}{8}$ inch (35.225 mm) in width. This usually refers to pipe with a 6-inch (154.2-mm) O.D. Although the coupons as torch cut or milled from the pipe have sharp edges, there is an allowable correction. The edges may be rounded. Some specifications even use the word *shall*.

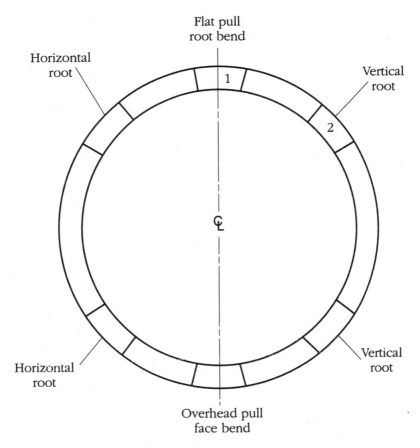

Flat pull
root bend

Horizontal
root

Vertical
root

1

2

Horizontal
root

Vertical
root

Overhead pull
face bend

12-7 *The layout of a pipe for the removal of test specimens for welder qualification.*

Even the boiler and pressure vessel code (ASME) does not consider a small crack on a knife edge of the coupon a flaw. Take advantage of all options.

Use a lightweight machine oil to lubricate the dies used in the guided bend process to cut down on drag as the horseshoe shape is formed. When completed for inspection, the legs of the specimen should be parallel to each other and of the same length. The weld material should be at the center of the radius bend. The most difficult test encountered is for water wall and superheater tubes in fired pressure vessels. Very little is printed on the requirements of individual companies. The companies must hold the proper stamp and the use of that stamp. They also must be bonded and carry insurance sufficient to cover all contingencies. They set their own test requirements.

One such test is set using 3-inch (77.1-mm) I.D. water wall tube with the same A number as will be used on the job. The short sections will be jigged in the 6G position. The root pass requires the use of the GTAW process. The tack welds should be *feathered*, or ground at both the start and crater end of each tack. When the root pass is finished, the subsequent passes are made using $\frac{3}{32}$-inch (2.3812-mm) low-hydrogen electrodes. All slag and any noticeable defects are ground to bright metal before any more welding is done. When the welding processes are complete, the procedure is continued as follows: The tube is first split in half and then quartered. The knife edges are rounded, and each quarter section is subjected to the guided bend test. There is no discard for starts and stops. The entire tube must be weld-perfect. In all of the other guided bend tests you simply never start or stop in an area that is designated for use as a coupon.

You may be told by uninformed people that there are only *qualified* welders, and *certified* welders don't exist. The real difference between the two is that the *certified* welder has the papers or documents to prove she or he is qualified to work with certain materials processes and procedures.

Welding regulation number 1 was put into effect by the State of Oregon in 1954. The following written material identifies the scope of the code for welding on low-pressure boilers. The salient parts of the code are still in effect. The welding processes now include GTAW and GMAW. Although not mentioned, the ASME code did include oxyacetylene welding tests on 4-inch (102.8-mm) O.D. schedule 40 black pipe. Section 9 of the Oregon pressure vessel code is used for fired pressure vessels now. The following regulations are still in use:

> *To: All Registered Mechanical Engineers, Plumbing and Heating Firms, Boilermaker Business Agents and Secretaries of Steamfitters Locals in Oregon*
>
> *From: F.W. Smith, Chief*
>
> *Boiler and Pressure Vessel Division, Salem, Oregon*
>
> **WELDING REGULATION NO. 1**
>
> *The state of Oregon has adopted all sections of the American Society of Mechanical Engineers Boiler and Pressure Vessel Code. Section IV of that code deals primarily with the construction of low pressure heating boilers and their safety appliances. We sincerely trust that the mechanical engineers who design heating plants and the contractors who install them will purchase Sec. IV of the ASME Code and become thoroughly familiar with its contents.*

The ASME low pressure heating boiler section makes it mandatory that welding on low pressure steel heating boilers be performed by certified welders and this department requires welding on the valves, pipes and fittings connected to such boilers to be done by certified welders. Furthermore, this department requires a hydrostatic test of not less than two times the stamped working pressure of hot water heating boilers to be applied to the welded pipes and fittings attached to such boilers prior to insulating the welded pipes and fittings.

The following rules have been formulated to afford reasonably certain protection to life and property and to provide a margin for deterioration in service. As the primary object of these rules is safety, the interest of engineers, contractors, welders and Steamfitters is of paramount importance.

These rules apply to welder performance for manual arc welding applied to pipe and fittings used in steam or hot water heating systems connected to boilers to be operated at pressures not exceeding 15 psi and hot water boilers to be operated at pressures not exceeding 160 psi or temperatures not exceeding 250 degrees F.

In an attempt to reduce the cost of material as well as welders' and inspectors' time, we have prepared the attached print to be used by firms and their welders as a guide in setting up for the test, which will determine without question, the ability of a person to perform sound welds.

1. A welder previously certified by this department to perform manual arc welding in all positions on pipe and fittings without the use of a backing strip need not be tested in accordance with the rules contained herein.

2. A welder certified without the use of a backing strip will be permitted to weld with or without the use of backing strip. A welder certified to weld with a backing strip only will not be permitted to weld without a backing strip until he/she passes the test in that procedure.

3. This test is to be made without the use of a backing strip unless the engineer or contractor demand the test to be made with a backing strip.

4. The new 6 inch or 8 inch Sched. 40, Spec. SA-53 black pipe. Two pieces are required, each piece 4 inches long with one end of each piece of pipe machine-cut to a 30 degree bevel leaving a $^1/_{16}$-inch land at the root of the bevel. Space the beveled ends not more than $^1/_8$ inch. Then place *the specimen in a horizontal position as shown under test position 5-G, lower right hand corner of print.*

5. *Start weld in center of discard coupon marked S under Figure Q-13.2 (A) and weld counter-clockwise to center of discard coupon marked F. If started at center of S, terminate weld at center of F and vice versa.*

6. *You will note Figure Q-13.2 (A) shows the order of removal of test specimens from welded pipe $\frac{1}{16}$ to $\frac{3}{4}$ inch in thickness. We shall use this figure for it is doubtful if pipe over $\frac{3}{4}$ inch thick will be used in low pressure heating systems. The above-mentioned figure shows six coupons in all, two root bend, two reduced section and two face bend. We shall not require the machining of the reduced section coupons unless the engineer or contractor so specifies. If not specified, the reduced section test coupons will be treated as root bend specimens.*

7. *Before welding is started, the inspector will stamp the welder's initials and coupon numbers so coupons can be identified when tested. The six coupons may be removed by flame cutting, then the edges must be ground smooth and the weld metal ground off flush with the surface of the parent metal. The coupons must be finished to $1\frac{3}{4}$ inch wide with a maximum tolerance of $\frac{1}{16}$ inch.*

8. *The firm or person who requests the test to be given will be billed for the service of our inspector in witnessing the welding and testing coupons. If the welder passes, both the firm and welder will receive a copy of the certification. If he does not pass, they will be notified accordingly and the test coupons may be examined by the welder within thirty days after the coupons are tested.*

9. *A welder who fails the initial test will be given a re-test when he feels that he is capable of passing. If not more than one coupon in the original test fails the welder may be retested immediately in the position in which he failed; but he must prepare two coupons in the same position as the one that failed.*

10. *A welder who passes the above described test will be certified to perform manual arc welding in all positions on pipe and fittings anywhere in the state of Oregon regardless of the name of the employer.*

11. *Should an engineer or a firm require a welder to be tested for high pressure piping the same procedure can* be used except for the thickness and diameter of the pipe to be used in the test and the removal of coupons.

12. If the high pressure piping test if given, use 8 or 10 inches new Sched. 40, Spec. SA-53 pipe and the coupons must be removed in the order shown under figure Q-13.2(b). If such a procedure is used the reduced section coupons must be machined by milling, especially at the reduced section area.

13. Our inspector will give full instructions regarding the machining of the reduced section coupons at the time the test is given. Should a contractor or welder use plate instead of pipe our inspector will explain the preparation of the plate specimens as shown on the attached print.

14. A welder who passes the test in high pressure piping will be permitted to perform manual arc welding on both high pressure and low pressure piping systems, except on high pressure pipe, headers and fittings that come within the scope of the ASME power boiler code. Welding on such construction must be done by a person who is certified in the procedures used by the individual or manufacturer who possesses and has authority to use the ASME code pressure piping symbol stamp.

15. FIELD WELDING: It will be to the advantage of the contractor, welder and all concerned if the ends to be joined by welding are properly beveled before weld metal is deposited. Proper fit-up of the joints to be welded will permit the welder to lay down the welding in the manner in which he was tested for it is practically impossible for a welder to obtain full penetration if roughly-cut joints are butted together without beveling.

The fees for the welding test are $15.00 per hour for our inspector's time witnessing the welding, and $1.00 for testing each coupon.

FIELD HYDROSTATIC TEST:

This test shall be applied and witnessed by an employee of the installation contractor or by an authorized state or insurance company boiler inspector.

This office will furnish free of charge our Form-B-7 which must be filled out by the contractor and signed by the welder or welders and the person who applied and witnessed the hydrostatic test. The completed form must be mailed to this *office within seventy-two (72) hours after the test or tests have been completed.*

Should the installation contractor or the owner or owners of the boiler or boilers to which the welded pipes and fitting are connected request the hydrostatic test to be witnessed by a boiler inspector employed by this department, the charge will be $15.00 per hour for our inspector's services while on the job witnessing the test, plus $15.00 per hour driving time and $.17 per mile for automobile expense provided the inspector is asked to make a special trip.

Welded joints between the main stop valve or valves and welded joints between the auxiliary stop valve or valves shall be subjected to a hydrostatic test in the presence of an authorized state or insurance company boiler and pressure vessel inspector for such welding comes within the scope of the ASME boiler and pressure vessel code.

The copies of certification papers as shown in Figs. 12-8 and 12-9 are typical examples of the types issued. The statement contained in item 1 of Welding Regulation No. 1 is wide-ranging. If you have the pressure-vessel certification papers and you continue to weld on pressure vessels and piping, state inspectors will honor the old papers or pocket card. Any inspector can require a retest if flaws are evident by visual inspection or NDT results.

The Army Corps of Engineers also honors certification papers. Personnel who had worked on penstocks for dams were sought for rip-rap work on rivers. Barge repair for river test equipment was also accepted without retest.

The welding, fabrication, installation, and repair of safety protection for heavy-equipment operators is another case in point. Many states require that all welding done on power shovels, road clearing, and grading housing (canopies) structures be done by certified welders.

All persons holding pressure-vessel certification papers need not retest. This is true for welding on fabricated compressed air reserve tanks. In most states, the tanks are inspected and tested before being put into service. An identification plate is attached. This plate is stamped and dated. Any welding repair must be done by certified welders, and the date or dates must be stamped on the plate. The tanks are inspected and dated each year.

Every general contractor involved in the types of work mentioned have their own test requirements. The procedures for weld testing,

BUREAU OF LABOR
STATE OF OREGON

September 11, 1957

Date of Test

This is to certify that .... John W. Shuster ....

is qualified to perform .... Manual Arc ....

welding in all positions with or without a backing ring on valves
pipes and fittings installed beyond the main and/or auxiliary
stop valves.

(Subject to provisions on reverse side)

Chief Boiler
Inspector

Commissioner

BUREAU OF LABOR
STATE OF OREGON

March 4, 1970

Date of Test

This is to certify that .... JOHN SHUSTER ....

is qualified to perform .... TIG ....

welding in all positions with or without a backing ring on
valves, pipes and fittings installed beyond the main and/or
auxiliary stop valves.   Max. thickness 5/8" P-1 Ma

(Subject to provisions on reverse side)

Chief Boiler
Inspector

Commissioner

12-8 *State "pocket" cards used to verify welder
certification to both contractors and inspectors.*

both foreign and domestic, are accepted, possibly because of insurance and bonding contracts. Except in special circumstances, one company will accept the papers of another company. There is often a time limit of 90 days on such acceptance.

The API code for welding on pipelines here and abroad is widely recognized. Since most pipelines involve low pressures, the code is not as restrictive as boiler and pressure-vessel codes. For pipes less than $2\frac{3}{8}$ inch (60.3 mm) in diameter, one root bend and one nick break test is required. As the pipe diameter increases, the tests are more in line with pressure vessel tests. The wall thickness of the pipe dictates the

CHICAGO BRIDGE & IRON COMPANY
WELDERS QUALIFICATION TEST
In Accordance With Section IX of the ASME Code - Latest Edition
FOR LOW HYDROGEN ELECTRODES ONLY

MATERIAL (PL or PIPE) _Plate_   FILLER METAL (F#) F- _4_

MATERIAL SPECIFICATION _A283 C_ TO _A283 C_   FILLER METAL (SA) SPECIFICATION SA- _SFH-5.1_

OF P NO. _1_ TO _1_

MATERIAL THICKNESS _7/8_   FILLER METAL (A#) A- _1_

		TWO SIDE BENDS
BACK GOUGE TO CLEAN METAL AND WELD OVERHEAD	3/8" ... 1/4" ... 7/8" ... 45°	Fig Q-7.1
	USE 1/8" ⌀ ELECTRODE, FIRST PASS BOTH SIDES. REMAINDER 5/32" ⌀ ELECTRODE.	RESULT 1 [OK]
	LAST LAYER MAY BE A SINGLE WEAVE PASS OR MADE WITH SEVERAL STRINGER BEADS	X-Ray
	THIS TEST QUALIFIES RANGE 3/16" TO MAX. TO BE WELDED.	RESULT 2 [ ]
OVERHEAD – BOTH SIDES		

		TWO SIDE BENDS
BACK GOUGE TO CLEAN METAL AND WELD HORIZONTAL	7/8" ... 3/16" ... 45° ... 1/8"	Fig Q-7.1
	USE 5/32" ⌀ ELECTRODE, FIRST AND LAST PASS, BOTH SIDES. REMAINDER 3/16" ⌀ ELECTRODE.	RESULT 1 [OK]
	THIS TEST QUALIFIES RANGE 3/16" TO MAX. TO BE WELDED.	X-Ray / RESULT 2 [ ]
HORIZONTAL – BOTH SIDES		

		TWO SIDE BENDS
BACK GOUGE TO CLEAN METAL AND WELD VERTICAL	3/16 ... 1/4" ... 7/8" ... 45°	Fig Q-7.1
	USE ALL 5/32" ⌀ ELECTRODES	RESULT 1 [OK]
	ALL PASSES UPHILL EXCEPT FIRST AND WASH PASSES WHICH MAY BE RUN DOWNHILL.	X-Ray
	THIS TEST QUALIFIES RANGE 3/16" TO MAX. TO BE WELDED.	RESULT 2 [ ]
VERTICAL – BOTH SIDES		

		TWO SIDE BENDS
BACK GOUGE TO CLEAN METAL AND WELD VERTICAL	1/8" ... 11/32" ... 45°	Fig Q-7.1
	USE ALL 1/8" ELECTRODE. ALL PASSES ARE TO BE RUN DOWNHILL.	RESULT 1 [OK]
	THIS TEST QUALIFIES RANGE 1/16" TO 3/4".	X-Ray
¹¹/₃₂" SINGLE BEVEL BUTT VERTICAL		RESULT 2 [ ]

1. QUALIFICATION ON BUTT WELDS ALSO QUALIFIES OPERATOR FOR FILLET WELDS.
2. QUALIFICATION ON F-4 ELECTRODES' QUALIFIES WELDER TO USE F-3, F-2 AND F-1 ELECTRODES.

PLATES TESTED	DATE	LOCATION	SOCIAL SECURITY NO.	BIRTH DATE	STARTED CB&I	SPECIMEN MARK
4	1974 July 23	Riddle Or		10-12-25	Year 1963	Jus S.

WE CERTIFY THAT THE STATEMENTS MADE IN THIS RECORD ARE CORRECT AND THAT THE TEST WELDS WERE PREPARED, WELDED AND TESTED IN ACCORDANCE WITH SECTION IX OF THE ASME CODE — LATEST EDITION.

CHICAGO BRIDGE & IRON COMPANY

BY _Wm H S Brown_

CB&I REPRESENTATIVE	FIRST	MIDDLE	LAST
	John	W	SHuster

12-9 *This type of certification paper is often honored by other contractors. It is almost always accepted for 90 days and sometimes for a year.*

use of root, face, or side bend tests. One factor that should have been mentioned previously is that the yield point and ultimate tensile strength should relate to the parent metal unless high-strength steel is involved and low yield point and ultimate strength values are stated.

The welds should never show evidence of failure or potential failure. These factors always relate to pull tests. Unlike most tests, the weld reinforcements of pipeline test coupons are not removed. This applies to root melt-through and any weld deposit showing above the outside diameter of the pipe. This relates to *nick break* and *tensile tests* only.

The guided bend tests do need the face and root ground smooth before testing. The specimens for bend tests shall have radiused, rounded corners. The allowance of a ⅛-inch (3.17-mm) defect or crack in any coupon gives the welder an excellent chance of passing the test. The radiographic films of girth welds in pipes with a ½-inch (12.7-mm) wall thickness gives some idea of acceptable standards. Slag inclusions shall not exceed ⅛ inch (3.17 mm) in width and the total length of separated inclusions shall not exceed ½ inch (12.7 mm) for any 12 inches (304.8 mm) of joint (seam) weld. The amount of dispersed or even inline porosity allowed is quite remarkable. That in no way affects the serviceability of the pipeline.

The general public is well served by the American Petroleum Institute. Its standards are not only recognized, but also used worldwide.

The tests employed by the Army Corps of Engineers are even more exacting. While the general contractors are responsible for the quality of everything used, the engineering inspectors are the final arbiters. They use all the test methods available, which includes all those previously mentioned and others, such as core samples, fracture toughness testing, even the removal of any suspect weld deposit and onsite inspection during that procedure.

The core sample tests are very small in relation to those done on the deposit of rock and concrete products. On fabricated and welded steel, the core sample may only be ½ inch (12.7 mm) in diameter and the length needed for the full depth of the weld root. Visual testing is first employed, then a dye penetrate and magnetic particle test. The Corps may require an acid etch and microscopic examination of the sample. The repair of core sampling has proved to be a real problem on bridge structures since the depth of core may exceed 2 inches (50.8 mm). The inclusion of slag and lack of fusion in this small area are distinct drawbacks. Fracture toughness is tested by using specimens of the size used in welding qualifications tests. They are ground to rectangular shape, and these tests are sometimes used for auxiliary piping as well penstock work.

The HAZ is the suspect area. Again, the etch test and a hardness test show the point or points for testing. The Charpy and fatigue crack test notch will be centered in the hardest area. In the case of carbide precipitation to grain boundaries, this area may be at the edge in the fusion zone. Any crack probably will not occur in the HAZ.

The arc air removal of suspect material has proven very satisfactory, especially with welding repair. The skilled operator can find the tiny gas pocket (porosity) or slag inclusion. The metal removed is on an incline at each end and only to the depth needed to remove the problem particle or void. Hopefully, the ditch will be from 6 to 8 inches (152.4 to 203.2 mm) or less in length.

On one penstock, 250 feet (approximately 77 meters) of weld deposit were removed from retainer rings not fused to the O.D. of the tube. The bevels as specified and used were of ample degree to allow complete root fusion. The Army Corps inspector noted welders *bridging* the bevel $\frac{1}{4}$ inch (6.35 mm) up from the root. The cost of repair for the contractor was much more than making the welds correctly the first time.

The first dye penetrant test probably was almost accidental. The first tests used powdered chalk from old doll bodies and kerosene. A crack was evidenced by a slightly darker yellow color than the surrounding metal.

Ultrasonic inspections can detect weld discontinuities that other methods may miss, including both surface and subsurface defects. It is an excellent way to check for laminations in materials as well as weld defects. The manner in which the system works is based on certain facts. Changes in wavelengths are reflected back. The waves are pulsed to an on/off pattern. The equipment first sends the wave then turns off and receives back some portion of the vibrational energy that was sent. Most commercial units use a video transmission system.

When voltage is applied to a quartz crystal, it vibrates rapidly, and it imports mechanic vibrations of like frequency to materials it contacts. Only a small part of the waves go into the material. If a film of oil is spread over the surface so that the crystal in the probe is joined to the material without the wave trying to pass through air, the energy of the wave is saved. Only about 5 percent of the energy will be transmitted into the steel through oil, and only 1 percent is present in air. The oil acts as a couplant, attaching the probe directly to the steel.

Medical labs use very similar equipment to pinpoint changes in arteries (wall thickness and narrowing or branching). The probe and couplant do the same jobs as on steel, but it is better because the tissue is not nearly so dense.

Figure 12-10 shows a typical pattern that would be transmitted and shown on an oscilloscope. The spikes are meant to show relative values. The type of equipment used may govern your test procedures. Most portable equipment has one transducer and a small screen. This screen is very similar to a 9-inch (228.6-mm) TV screen. The probe, shown as a small rectangle, has the numbers 1, 2, 3, and 4 inside. It may have an angle transducer, which means that the beam of energy strikes the steel or weld at an angle rather than straight down.

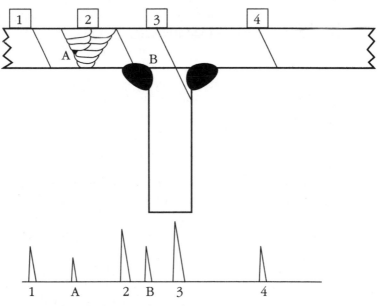

#1   Shows thickness of metal

"A"   Shows weld flaw

#2   Penetration into whole weld

"B"   Incomplete tie at root

#3   Very dense material

#4   Again shows top plate thickness

12-10 *A simulation of wave patterns as projected on a video screen.*

It takes some practice to pinpoint a defect using this equipment. Some testing labs use two transducers. One sends and the other receives the energy. Tape-recording devices are now available to make a permanent record of the findings. One problem is evident in Fig. 12-10: spikes 1, A, and 4 are almost identical in length, which means that the skilled operator must evaluate *all* metal thickness and possible areas of defect.

The use of ultrasonics has been of extreme value to the boiler industries. All of the large companies both install and service their products. The companies owning the boilers are constantly checking the condition of all equipment. At regular intervals the superheater tubes are removed and replaced, a process called a *turnaround.* The larger water wall tubes on the sides of the heating chamber are replaced only if a possible thinning or weakness of material is found or suspected. The skilled ultrasonic technician can find evidence of such thinning by measuring the wave spike of a new tube thickness against that of a tube wall that has been in service for some time.

There are often mineral deposits inside the tubes, so care must be taken to evaluate this factor. If a skilled operator fine-tunes a *scope* calibrated for plate thickness, the near exact depth of a flaw can be pinpointed, but size may be a factor. Complete information on wave velocities in steel relative to energy input and through which media are available in table form from the company supplying the equipment.

Remember that waves, as they enter steel, do not travel in a straight line. If you wade into a clear pool of water and look down at your legs, they appear to bend and go off at an unnatural angle. So do waves in steel bend so that a probe is not directly in line with a flaw. We must learn to find the flaw depending on the couplant (oil, water, etc.), the thickness of the steel, and its density. By training oneself with standard equipment and these variables you will soon become adept at locating problem areas. Short rays travel along pipe walls for considerable distances.

Radiographic testing

The radioactive isotopes are cobalt-60, indium-192, thulium-170, and cesium-137. These isotopes are usually contained in lead casing or sometimes spent uranium vaults, if needed. They pose real hazards to operators. A U.S. license is required for the use of these materials.

The image produced in every case is the opposite of photography. The film is placed behind the steel, and the penetrating rays produced by the X-ray cathode tube pass through the steel and develop the film. The denser metal absorbs more energy, and cracks and voids absorb less energy, producing dark area on the film. Porosity generally shows a dark outline of the flaw.

The amplified electrons produced by the generating unit strike a tungsten target and increase in speed through the tube into the air and into the steel. Portable X-ray units produce a wavelength 1/10,000 that of visible light. High-voltage generators produce 1 to 2000 kilovolts and will penetrate 5 to 9 inches of steel (127 to 228.6 mm).

When we speak of the pictures produced, we lump them as radiographs, but those made using isotopes produce a shorter wavelength take longer to develop the film and should really be called gammagraphs because they are produced by gamma rays caused by slow disintegration of the radioactive isotopes.

The cost of radiographic tests limit their use. The equipment needed is expensive. The minimum two-year training period for the licensing of inspectors is a factor. Since most companies must hire outside testing agencies, travel expenses and the actual inspection fees seem high.

Do a thorough cost analysis before bidding on any fabricated and welded structure that requires this type of test. Make sure that the inspection team's schedule will match the date of your inspection needs.

One last test should be mentioned. It is very cost effective and anyone can use it. The soapsuds test is simple. Any closed system can be checked for leaks. Compressed air is cheap, and if no convenient fitting is already in the system, a threaded fitting is easy to install. It works just as the air intake valve on your automobile tires. If the air hose has a pressure gauge, a few pounds of pressure is all that is needed. Never use high pressure when checking for leaks. A pressure of 10 psi (7 kilograms per square centimeter) is sufficient. High pressure may not damage the system, but it will blow bubbles away before a leak can be detected.

Any high-sudsing soap is fine for use. Mix with enough water to completely cover any suspect area. If your gauge pressure is now lower than what you first used, add a little more air. This fact alone probably indicates a leak. A small paintbrush is fine for brushing the suds onto steel. If the bubbles increase in size or many small bubbles appear in one area or seam, mark the spot for repair. It might be wise to wipe the area dry and again apply suds. This test is very safe for oxy-fuel tanks, lines, hoses, connections, torches, or other related equipment. Never use oxygen in place of compressed air.

Any container, line, or system can be checked in this manner. There is no danger from this test, even with volatile fuels. Sometimes nitrogen is introduced into lines carrying telephone and electrical power cable. Nitrogen acts as a drying and moisture-retarding agent. Soapsuds will not damage the plastic tubes or pipes. The low pressure involved makes it a fine agent in the leak-testing applications. Of course, hoses and small containers capable of holding air pressure can be immersed in still, clear water. If bubbles appear, a leak is evident.

Cooperation

Many of the codes for steels and welding have been previously covered. The major code-making bodies that have not been mentioned are Aerospace Industries Association of America (AIAA), American Society for Metals (ASM), American Society for Nondestructive Testing (ASNT), American Water Works Association (AWWA), Mechanical Contractors Association of America (MCAA), National Board of Boiler and Pressure Vessel Inspectors (NBBPVI), National Certified Pipe Welding Bureau (NCPWB), Pipe Fabrication Institute (PFI), Steel Plate Fabricators Institute (SPFI), Steel Plate Fabricators Association (SPFA), and the Welding Research Council.

One complaint is noted nationwide. Most contractors and fabrication company supervisors have little knowledge of steel specifications. They often don't have a welding procedure process. Even process data is ignored. Inspection is a painful problem, and it costs money. When welds are rejected by an inspection firm, the inspector blames the welders and then state that the AWS code or codes are not followed. The contractors and field erectors don't follow procedures. They don't keep records of welder qualifications. They don't understand nondestructive test methods. Inspectors say that 50 to 75 percent of rejection is because someone did not understand weld symbols. They also blame the fabricators for not maintaining steel records and not asking inspectors to check the stock. This type of record-keeping is expensive. Lot, batch, and heat numbers from steel shipments may be difficult to obtain. As the codes become more restrictive, and the inspectors want more respect, they may have to earn it.

The fabrication and erection companies have been in business for years. There have been few failures. If the codes are there to protect inspectors' jobs, then the companies cannot support the expense and go out of business. The inspectors produce nothing but *safe practices*, which is difficult to sell to a customer.

A little more cooperation is the answer. A visit to a company by an inspector may not make the inspector any money, but an offer of help on future problems may save both companies. Maybe we need to simplify codes. Section 9 of the ASME boiler code is a case in point.

The AWS D1.1 may or may not reflect this, but it depends on your view of the situation. This text answers and addresses many of the problems involved. Study the weld symbol and the explanation of its use. The AWS was kind enough to permit its inclusion. Use the chapter on procedures and requirements to help you and your company work toward a better fabricated and welded product.

The explanation of fabrication practices should be a real help. The tricks of the trade will save time and money. The terms used in the industry will be a signal to others that you do understand the fabricator-welder language. A new OSHA rule has just been put into effect that requires a full body harness on field erection jobs where above-ground-level work is involved. Study the chapter on safety as though your life depended on it.

Project 10: Welder qualification test

Make up and weld the test plates for welder qualification tests as shown in Figs. 12-3 or 12-4. The use of a backing plate is optional. Add a strike plate of ⅜- × -2- × -2-inch (9.525- × -50.4- × -50.4-mm) material to the weld start point. Add a second like-sized plate to the end of the test plate. These small square plates are often called *run-in* and *run-off* tabs. They usually are grooved where they meet the included angle of the bevel so that you will not false-start your weld on the side of the bevel on your test plate. It also supplies preheat to the major plate area. It provides real confidence that you will attain complete melt-through (weld reinforcement) at the root. The run-off tab at the other end eliminates the worry of crater cracking. This is a way to ensure complete weld integrity if a test specification calls for no discard material and coupons taken from plate edges. The tabs are removed when the welding is completed. Tabs are generally used for submerged arc welding to eliminate lack of fusion.

You might want to use a 35-degree bevel on each plate. This bevel is permissible in most cases. If you choose not to use a backing plate, you may need a small land ground on the knife edge of the bevel. You will be right back to two 30-degree angles if you do. Remove the backing when the welding is complete if any is used. Inspect the weld area. If an inspector needs to see the test plate before the coupons are removed, be sure this is done. A track torch is a good way to remove them. Grind the edges so that there is no chance of a crack starting in a rough spot. Round them if there is no restriction.

If you do not have a mandrel and die, improvise. Any hydraulic press or jack can be used. A short piece of shafting of the right size becomes a mandrel. Any scrap of H beam leaves enough space to start the bend. Once the radius starts to form, the rest is simple. Use the jack or press to make the sides parallel as shown in Fig. 12-5.

The use of a hammer is not recommended. The force used should be as uniform as possible. Hammer marks, particularly on coupon edges, can start cracks. Indentations may be mistaken for internal flaws. If you are satisfied with your visual inspection of the coupons, they are ready for review or qualification.

Questions for study

1. What causes bolt or rivet failure?
2. What type of welding process was used on many of the early bridge construction projects?
3. Why is rain a problem for welders?
4. What do the letters HAZ mean?
5. How do you remove the excess crown on a deep groove weld?
6. What are wagon tracks?
7. Name five societies or rule-making bodies involved in the quality production of steels.
8. What types of information can such a body supply for a given steel number?
9. What is a macroetch test?
10. Can you explain how a magnetic particle test works?
11. How can you make a quick check on the web thickness of channels and beams?
12. How many rule-making bodies license fabricators?
13. In this book, where would you find information on detailed procedures for welding methods, joint details, materials, personal qualifications, and other related information?
14. What are the G numbers, and what do they mean?
15. In Fig. 12-3, how many root bend test specimens are needed for welder qualification?
16. On a guided bend test, where are any flaws likely to appear?
17. What is the danger of changing electrode angles when using low-hydrogen rods in the vertical position?
18. How can you correct for an included bevel angle that is too small in a groove weld?

19. On welder qualification guided bend tests, you may use
   _____ to lubricate the mandrel and dies.
20. What rule-making body or society has set the rules for boiler
   fabrication and welding?
21. How are the boiler and pressure vessel test coupons
   identified for welder qualification?
22. What kind of test is used to qualify boilers for use?
23. If a welder has certification papers for pipe welding (low
   pressure), what other firms or inspectors will accept the
   papers without retest?
24. When a section of a welder qualification coupon is reduced
   in width, what type of test will be used on that specimen or
   specimens?
25. In what way are pipeline tests for welder qualification less
   restrictive than pressure vessel tests?
26. What problems are encountered in the repair of steel
   structures where small core samples have been taken?
27. In ultrasonic testing, what percentage of wave energy actually
   enters the steel?
28. How many transducers are usually employed for portable
   ultrasonic test equipment?
29. What is a *turnaround* in boiler terminology?
30. In radiographic tests, where is the film placed for exposure?
31. Why are radiographic tests more expensive than most
   other tests?
32. When testing for leaks in closed containers, hoses, fittings,
   and tanks, what materials can be used for low-cost testing?
33. How much pressure is used for this low-cost test procedure?

Abbreviations

AASH	American Association of Street Highway Transportation
ABS	American Bureau of Shipping
ac	Alternating current
AISI	American Iron and Steel Institute
ANSI	American National Standards Institute
API	American Petroleum Institute
APPD	Approved
ASME	American Society of Mechanical Engineers
ASSY	Assembly
ASTM	American Society for Testing and Materials
AWG	American Wire Gauge
AV	Arc voltage
AWS	American Welding Society
BC	Bolt circle; between centers
BE	Beveled end
BRS	Brass
BL-P-PE	Black pipe, plain end
BOM	Bill of materials
B&S	Brown and Sharp
BW	Butt weld
BWG	Birmingham Wire Gauge
C	Channel steel
CFM	Cubic feet per minute
CHG	Change
CKS	Countersink
CI	Cast iron
CL	Centerline
CON	Concentric
CS	Carbon steel; cold spring
CV	Constant voltage
CYL	Cylinder
dc	Direct current
DCE	Direct current electrode negative
DCEP	Direct current electrode positive
DIA	Diameter
DT	Destructive testing
DWG	Drawing
ECC	Eccentric

EL	Elevation
ELEV	Elevation
ELL	Elbow
ELON	Elongation
ERW	Electric resistance weld
ESW	Electroslag welding
EXIST	Existing
FAB	Fabricate
FCAW	Flux core arc welding
F-F	Face to face
FLG	Flange
FOB	Flat on bottom; free on board
FOT	Flat on top
FOW	Forge welding
FRP	Fiberglass reinforced pipe
FW	Field weld
GMAW	Gas melalarc welding (mig welding)
gph	Gallons per hour
gpm	Gallons per minute
GTAW	Gas tungsten arc welding
HAZ	Heat-affected zone
HEX	Hexagonal
I.D.	Inside diameter
IN	Inches
ISA	Instrument Society of America
ISO	Isometric
L	Angle steel
LH	Left hand
LR	Long radius
MAX	Maximum
MIN	Minimum
MI	Malleable iron
MIL	Military or federal spec
MS	Mild steel
NC	National fine threads
NDT	Nondestructive testing
NPT	National pipe thread
OAW	Oxyacetylene welding
O.D.	Outside diameter
PEPL	Plain end, no thread plate
pcs	Pieces
RAD	Radius

psi	Pounds per square inch
QWP	Qualified welding procedure
QQ	Federal specification
REF	Reference
RH	Right hand
rpm	Revolutions per minute
RV	Revision
SAE	Society of Automotive Engineers
SAW	Submerged arc welding
Sch	Schedule
Sch 40	Standard wall pipe
Sch 80	Double-wall thickness
Sch 120	Triple-wall thickness
SMAW	Shielded metal arc welding
Spec	Specification
SPFA	Steel Plate Fabricator Association
SR	Short radius
SS	Stainless steel
STD	Standard
STL	Steel, also MS
SW	Stud welding
SYL	Symmetrical
SYMM	Symmetrical
TIG	GTAW, gas tungsten arc welding
TK	Tank
TL	Top level
TOS	Top of steel
TS	Tensile strength
TYP	Typical
U.L.	Underwriters Laboratories Inc.
UM-PL	Universal mill plate
W	Wide flange steel shape
WI	Wrought iron
WOG	Water, oil, and gas
YP	Yield point
XH	Extra-heavy pipe
XS	Extra-strong pipe
XXS	Double extra-strong pipe

Glossary

abrasion Wear from natural elements (water, wind, etc.) or metal-to-metal contact in movement.

alloys All elements added to steel.

amperage or **amps** The amount of current flowing in a circuit.

asbestos A silicate of calcium and magnesium once used as insulation. Asbestos is a known caustic agent of respiratory and lung disease and a cancer-causing material.

backing strip Any steel strip affixed to the back of a beveled open root or grooved joint.

backstep A welding procedure in which a short length of weld is made and then the welder skip ahead about the same length to start a bead and weld back until the bead ties in with the first bead end.

bid bonding Cost of guaranteeing job completion and other contract provisions.

blueprint Visual instructions of fabrication and welding procedures; a road map showing details in one-dimensional views of a finished product.

boilermakers Those person working with fired and unfired pressure vessels, boilers, compressed air tanks, etc.; all pipes and tubes above 20 inches (508 mm) in diameter where pressure is involved.

bonding A joining of surfaces usually by infiltration of the surface crystals by a filler metal compatible to both.

borrod A filler metal, usually rod or stick form. The weld deposit is of very hard tool-steel quality.

bottlenecking Narrowing avenues of production; loss of production by cutting access to the next stage of fabrication or welding.

bridging Welding the side walls of beveled or grooved plate or pipe; no attempt is made for root weld fusion.

butter passing A process in which a filler metal is used to coat the edges of separate pieces of steel and then the coated edges are welded together using that same type of filler metal.

283

cambering Arching of a steel shape or section.

carbide precipitation A process in which some alloys and carbon move to the grain boundaries (outside edges of the crystal structure) instead of staying evenly dispersed.

cherry picker A small mobile crane.

clips Short pieces of angle iron 3 to 4 inches (76.2 mm to 101.6 mm) long with a discretionary leg thickness. A drilled or stamped hole is made in the leg center. Wire rope, shackles, turnbuckles, or other tools may be attached through the hole or to bolts passed through the hole.

code A body of rules, regulations, or even laws covering practices of fabrication and welding.

constant potential Constant voltage without regard to the amp output.

Coolie hat A slightly domed top for a round tank; if this term is not acceptable, use pitch or rise and run figures from edge to center.

cored wire Filler metal with flux or alloys in a small steel tube, coiled for use in the gas metal arc process.

crater The area at the end of a weld bead; it may have a sunken appearance and is not fully protected by slag.

cross-keying Laying loose materials in layers so that the second layer is perpendicular to the first layer.

crystals The microscopic shape of steel similar to that found in breaking iron ore.

cutter One who cuts steel with a torch.

declivity board A tapered piece of material that reflects the true level of material that is placed on an incline.

diamond plate Flat steel plate with raised diamond formation shapes on the top surface.

dog or dogged Steel held in place by the use of clamps, strongbacks, or dogs and wedges.

downtime A period when machinery or equipment is not running and workers are idle.

drift pins Round tapered tools for alignment of bolt holes in fabricated structures.

electrodes Rods, usually coated, used as filler metals for the shielded metal arc welding process.

elongation factor The amount a steel will stretch before failure; generally measured as a percentage factor in a 2-inch section of steel.

facia Steel to cover and soften the look of raw steel; also ornamental iron steel.

field work The fabrication, welding, and erection of steel outside of a shop.

fingernailing A condition in which the welding electrode coat burns away more on one side of the rod than the other.

flange The outside surface or surfaces of a structural shape.

full tuck Strands of wire rope (cable) braided back into itself for strength and clean ends.

gas manifolds The series hookup of gas cylinders to provide large supplies of gas to equipment.

gilley-ga-hike Crane operators' name for a small mobile crane.

HAZ Heat-affected zone in steels out to a point where the steel does not reach a temperature of 200 degrees F during the welding process.

HY steels High yield point steel.

Hadfield manganese Steel with 16 to 18 percent of manganese added to the steel.

hard surfacing The fusion-welding of beads to a surface; the filler metal used depends on the type of wear that may be encountered.

harness package Complete light reflectors, wiring, and wire-covering materials for a towed vehicle. It should include four-way plug-type receptacles to match brake light and turn signal operation of the tow vehicle.

hot slab A massive steel table with openings for dogging devices to hold steel in place as bending and fabricating procedures are used.

hydrostatic test Often called a squeeze test; boiler tubes are filled with water and additional pressure is applied by pumps; the gauges are watches closely until the desired pressure is reached.

interpass temperature Usually the temperature of the steel at the edge of each bead during the welding process.

kerf The width of material removed when steel is cut with a torch.

killed steel Steel refined and deoxidized when melted to prevent gases from being included as it solidifies.

laminated steel Steel that is not solid but has spacing between two sheets of material. It is not acceptable but may be encountered in older plate or shapes.

land That part of plate or pipe not cut on an angle when the steel is chamfered rather than beveled. A land is also produced when the sharp edge of a bevel is blunted by grinding, machining, or filing.

martensitic Crystal structure of steel when treated to obtain maximum hardness.

melt-through Uniform, complete weld bead penetration through a bevel or groove joint required for a welder qualification without use of a backing strip.

metallurgy The study of how and why steels and other metals perform as they do.

miter cut Cutting at an angle of any shape, for example, a 45-degree cut.

Pythagorean theorem A geometrical theorem by a Greek mathematician that states that in any right triangle, the square of the hypotenuse is equal to the sum of the squares of the other two sides.

pattern loft Any area in which patterns are developed from blueprints or sketches.

penstock Tunnel liner or other large-diameter, thick-wall steel pipe for conveying water down to a power house.

pneumatic tools Tools that use compressed air for motive power.

positioner Any equipment that turns fabricated sections into a more favorable position for fitting or welding.

prefabrication Any work done before the first tack welds are started.

radius Half the diameter of a circle.

rectifier A device that changes ac current to dc current.

reverse polarity Electrode positive, ground negative; deep penetration, particularly with high-voltage factors involved (direct current electrode positive, or DCEP).

root pass The first weld bead in fillet or groove procedures.

rosebud A heating tip with many orifices for use with oxy-fuel torches.

scaled blueprint A blueprint that translates all dimensions to an exact scale. For example, ½ inch on a blueprint might equal 1 foot on the actual structure.

shipfitter A fabricator of steel and constructor of boats (ships).

shrinking Reducing both the size of crystal formation of steel and the air spaces between the crystals, usually by heating and applying cold water.

soak Reheating the steel until it can be worked.

spalling A condition that may occur when two or more layers of hard surfacing are put directly on top of each other without a "soft filler" metal layer in between. The hard material will stress-crack and tear away from the parent metal.

steamfitters Those persons working with pressure piping up to 20 inches (508 mm) in diameter, outside of boilers.

straight polarity Electrode negative, ground positive; not noted for deep penetration (direct current electrode negative, or DCEN).

stress relieving Heating steel to a required temperature and allowing it to cool slowly over a specific time period.

strut A brace usually used to support or strengthen to steel shapes or sections.

submerged arc welding (SAW) The arc is covered with powdered alloys or salt-free dried sand. The alloys become part of the weld material, and the other powders or sand become a protective slag cover.

superheater boiler tubes Steam tubes or pipes at the top of a boiler. These tubes carry the highest steam pressure. They are enclosed at the top of the furnace chamber. The trade name for the enclosure is *the penthouse.*

swaged A steel socket clamped onto wire rope to provide a smooth clean end.

sweep Arching in the horizontal plane, which is not usually desirable.

track burner A motorized unit that travels on track in conjunction with oxy-fuel or plasma arc torches for the cutting of steel plate.

transition Change of size or shape of steel to match existing fabricated items.

turnbuckle A coupling between lengths of wire rope or rod, usually loop ends or hook ends; these ends are threaded for a tightening action.

valley Steel plate formed with a slight V contour; used to carry rain water from adjoining sections of steel buildings.

voltage The force of the arc between the electrode tip and the weld puddle.

web The inside surface that attaches to and supports the flange or flanges of a structural shape.

weld flaw Any defect found in a weld deposit.

weld nugget A term used to denote the weld area from spot or resistance welding; used by welders to mean any area of weld material.

weld porosity Gas bubbles trapped in the weld bead.

weld specimen A coupon or strip of steel with a weld in the center used for test purposes.

welder certification Written and attested proof that the welder has passed certain tests.

welder qualification The passing of tests on proficiency covered by written specifications, procedures, and processes.

welding leads Power cables from the machine to the work and from the ground clamp back to the machine.

welding procedure A step-by-step instruction covering such factors as number of passes, filler metal type, process used, arc travel speed, current rating in amps and bolts, material to be joined, and shielding gases, if required.

welding process A method of welding—GTAW, GMAW, etc.

wiring A leveling process to match the surfaces of steel plates.

Index

Illustration page numbers are in **boldface**.

About the Author

John W. Shuster is a licensed boiler and pressure vessel mechanic and welder. He has served as a construction millwright and foreman for the building of sawmill and rock crushing equipment. For more than 10 years he operated M&S Metal Craft Company, a steel fabrication business that manufactured logging and sawmill equipment, rock-crushing equipment, and automatic resurfacing (of rock-crushers) equipment. Mr. Shuster has taught community-college classes in fabrication, metallurgy, drafting, blueprint reading, structural steel estimating, and all types of welding processes for over 22 years. He wrote the first course outline for an associate's degree in welding technology.